MADE NOT BORN

The Troubling World of Biotechnology

Edited by Casey Walker

Sierra Club Books
San Francisco

The Sierra Club, founded in 1892 by John Muir, has devoted itself to the study and protection of the Earth's scenic and ecological resources—mountains, wetlands, woodlands, wild shores and rivers, deserts and plains. The publishing program of the Sierra Club offers books to the public as a nonprofit educational service in the hope that they may enlarge the public's understanding of the Club's basic concerns. The point of view expressed in each book, however, does not necessarily represent that of the Club. The Sierra Club has some sixty chapters coast to coast, in Canada, Hawaii and Alaska. For information about how you may participate in its programs to preserve wilderness and the quality of life, please address inquiries to Sierra Club, 85 Second Street, San Francisco, CA 94105.

www.Sierra.org/books

The publisher gratefully acknowledges permission to reprint the following copyrighted texts: "Thy Life's a Miracle," from *Life Is A Miracle: An Essay Against Modern Superstition* by Wendell Berry, © 2000 by Wendell Berry, reprinted by permission of the Perseus Books Group, "Of Gadgeteers and Bionic Deer: The Biotech Bastardization of Hunting," from *Heartsblood* by David Petersen, © 2000 by David Petersen, published by Island Press, Washington, D.C., and Covelo, CA.

Published by Sierra Club Books, in conjunction with Crown Publishers, New York, New York.

Member of the Crown Publishing Group.

Random House, Inc., New York, Toronto, London, Sydney, Auckland

www.randomhouse.com

Printed in the United States of America on acid-free paper containing a minimum of 50% recovered waste paper, of which at least 10% of the fiber content is post-consumer waste.

Design by Leonard W. Henderson

Library of Congress Cataloging-in-Publication Data

Made not born : the troubling world of biotechnology / edited by Casey Walker.

p. cm.

Includes bibliographical references and index.

1. Biotechnology—Popular works. 2. Biotechnology industries—Popular works.

I. Walker, Casey.

TP248.215 .M33 2000

666.6—dc21 00-030104

ISBN 1-57805-059-6 (paper)

10 9 8 7 6 5 4 3 2 1

First Edition

For the inimitable wild ducks—Carolyn, Nicole, Debbie,
Brooke, Kristin, and Weston

Contents

Acknowledgments

The contents of this book originally appeared as *Wild Duck Review*'s issue on "Biotechnology" (vol. V, no. 2), which was made possible through the generous support of the Foundation for Deep Ecology, Norcross Wildlife Foundation, Lawson Valentine Foundation, HGH Foundation/Blue Mountain Center, Jenifer Altman Fund, Solidago Foundation, and Peradam Foundation, and through the support of many extraordinary individuals who work to hold the line for a "world of born."

My many thanks go to *Wild Duck Review*'s original contributors, who gave of their time and talents, proving true intelligence is always generous: Wendell Berry, Chris Desser, Stuart Newman, Jerry Martien, Marti Crouch, David Loy, Andrew Kimbrell, Jack Turner, Richard Hayes, Catherine Keller, Richard Strohman, Freeman House, Elizabeth Herron, Maya Rani Khosla, Kristin Dawkins, Martin Teitel, David Petersen, and Hank Meals. Special recognition and thanks go to Joanna Robinson, who labored to transcribe, proofread, and copyedit out of sheer will and love and belief in the work; and to Gary Clelan and John Girton for their technical wizardry.

Thanks, too, to the *Wild Duck Review*'s editorial advisers for adding that ineffable presence to what is surely the most improbable journal in today's media world: Will Baker, Chris Desser, Jim Dodge, Mark Dowie, Todd Gitlin, Robert Hass, Elizabeth Herron, Jane Hirshfield, Lewis Lapham, Joanna Macy, Jerry Mander, Jerry Martien, Hank Meals, David Noble, Joanna Robinson, Marilynne Robinson, Richard Rorty, Gary Snyder, and Charlene Spretnak.

Special thanks to editor Danny Moses at Sierra Club Books, who first suggested the idea of this book, and has, throughout the course of its making, cared for it as only an exceptional editor could. And, thanks to literary agent Victoria Shoemaker for advice both generously and gracefully given.

Finally, my deepest gratitude goes to my children—Brooke, Kristin, and Weston—whose daily company is beyond measure and beyond any telling.

Casey Walker, *Wild Duck Review* Editor & Publisher

Made Not Born

CASEY WALKER

TEN POINTS TO INTRODUCE BIOTECHNOLOGY

1. Biotechnology applies knowledge from the life sciences to the design and creation of living things. Consequences of bioengineering are consequences directly tied to each individual life and the whole of the living world. These consequences are of an order entirely different from the consequences of technologies such as automobiles, nuclear energy, or computers.
2. Transgenic biotechnology creates living things that would not be alive otherwise—salmon with human and chicken genes, tobacco with firefly genes, potatoes with pesticides—launching a new order of artifice that is distinctly different from the bits of paper, plastic, or aluminum we pick up as trash on the sides of roads and recycle. These new life forms will interact as living things do within the living world; they will give off pollen and roots, swim, fly, run, mate, multiply, eat, be eaten, die, and decompose. These inventions will be subject to the same systems of values and rights as all living things.
3. Biotechnology is the technology of an industry that, as is true of all industries, has as its sole purpose the making of money. Its power to make money rests on two conditions: that it be well-capitalized, and that it correctly predict robust markets. Today, the number and size of acquisitions, mergers, venture capital, and new upstarts of biotech companies and biotech complexes predict an industry that will someday exceed those of computer technologies and media/entertainment. Biotechnology's current and/or anticipated markets include: (1) genetic engineering for crops, livestock, fish-

eries, and forests; (2) genetic engineering for humans who are unborn, sick, maimed, or physically or mentally nonconforming; (3) genetic information useful to numerous databanks, including those of national and domestic security forces, criminal forensics, institutions of education and employment, and medical, life, and disability insurance companies; (4) the possibility of human clones as perpetual organ/limb/skin factories and genetically engineered chimeras (part human, part primate) as a new labor class; and (5) biowarfare and terrorism with products targeting genetic populations of crops, livestock, and people.

4. Biotechnology creates living things that will affect ecosystems in ways that cannot be recalled like faulty brakes, cannot be cleaned up like an oil spill, and cannot be stored like radioactive waste. Genetically engineered live viruses meant for influenza immunizations or for cancer treatments can mutate beyond targeted host cells and create new, epidemic viruses. The escape of genetically engineered crops, livestock, fish, or trees into wild populations can irrevocably mix with wild species, thus homogenizing the genetic materials of artificial, domestic, and wild species (a new form of extinction) and rendering those species vulnerable to single pathogens. Collateral effects to biodiversity losses from genetic "weediness" within any ecosystem are incalculable, with impacts extending from microbes, soils, insects, birds, and plants to oxygen levels, dew points, and weather patterns. Conversely, if current and future attempts to resurrect extinct species (such as with the Tasmanian tiger in Australia and the Huia bird in New Zealand) are successful—Jurassic Park–like—and each animal is subsequently introduced into zoos, parks, and wild ecosystems, then equally new and different ecosystemic impacts will occur. In all cases, the natural evolution of species and ecosystems will be corrupted and suffused with artifice.
5. Biotechnology as an industry is not self-correcting or self-regulating

and has little precautionary regulation or oversight in place. Moratoriums and bans on genetically engineered organisms have been difficult to impose and harder to maintain. International trade law to date has refused to recognize genetically engineered foods as different from nonengineered foods, and has mandated patent systems in every country to accommodate and enforce ownership of genetically engineered organisms. The United States is the world leader in genetically engineered organisms. The U.S. Department of Agriculture (USDA) is co-owner, with Delta Land & Pine/Monsanto, of terminator technology (which disables seeds from germination and obliges farmers to buy new seed each year). Industry officials expect U.S. agricultural exports to be 90 percent genetically engineered within a decade. In April 1999, President Bill Clinton awarded four Monsanto scientists the National Medal of Technology for the birth of agrotechnology and for placing the United States at the forefront of a new science. Thus far, food labeling and safety testing have not been required for the highest yielding, engineered crop and livestock by-products commonly consumed in milk, cheese, ice cream, eggs, meats, potatoes, tomatoes, corn, soya, fast-food burgers and french fries, corn and potato chips, and baby formulas and baby foods. Further, genetically engineered organisms are legally regarded as the intellectual property of their inventors through U.S. patent law. Throughout the seventeen- to twenty-year span of an awarded patent, inventors are ensured exclusive legal ownership for the commercial application of genetically engineered life forms.

6. Public criticism of biotechnology is, in the United States, virtually nonexistent. Public acceptance or support for biotechnologies is convergent with cultural beliefs that technological innovations are progressive, inevitable, and the best means to economically compete and succeed in global markets. In addition, the mediated world of a predominantly consumer/entertainment culture is convergent with

the spectacular, limits-defying feats promised by biotech. Public ignorance of ecological, moral, and human social issues presented by biotechnology can be attributed to: (1) de factor censorship through corporate intimidation and lawsuits; (2) self-censorship of career journalists and corporate-owned media; (3) aggressive public media campaigns by biotech industries; (4) contractual ties between biotech industries and universities, which tend to foreclose contrary research and contrary voices; (5) a failure in education to emphasize values, human beings, consciousness, questions, and conscience, as well as a lack of ecological literacy (preschool through university); (6) weak engagement of spiritual consciousness or practice; and (7) a politically disengaged public.

7. Advocates tend to characterize biogenetic engineering as problem solving. Critics tend to redescribe both the problem and the solution. One compelling problem has been identified as world hunger, with biotech advocates arguing for a "Second Green Revolution" to meet the needs of an estimated human population of ten to twelve billion in the twenty-first century. Critics argue that feeding hungry people is a distribution not a supply problem, a problem best solved, in any case, by small-scale agriculture with ecologically sustainable strategies that are independent of genetically engineered seeds, monocultures, and factory-style livestock. A second problem area covers health and medical issues ranging from conditions as simple as diarrhea in Third World countries to complex medical conditions such as diabetes and cancer. Critics argue that the root causes of many medical conditions are industry-related environmental toxins and contaminated air, water, and food supplies, for which bioengineering solutions are merely palliative. Beyond strict medical applications, there exists a strong cultural bias for the techno-eugenic elimination of human imperfections, inferiorities, aging, and even death through genetic engineering. One group, the Extropians, hopes to solve the problems of biological barriers and become postbiological, post-

humans: "persons of unprecedented physical, intellectual, and psychological ability, self-programming and self-defining, potentially immortal, unlimited individuals." An Extropian conference at the University of California-Berkeley in August 1999 featured such well-known scientists as Gregory Stock (UCLA), Cynthia Kenyon (UCSF), Calvin Harley (chief scientist at Geron), Eric Drexler, and Roy Wolford. Critics argue against ideas of biological imperfection and perfectibility as diminished and diminishing views of humanity and of life processes.

8. Biotech scientists, writers, and industry advocates frequently disregard criticism as ignorant, hysterical, or sentimental. Yet a growing number of cellular and molecular biologists (see Stuart Newman and Richard Strohman, herein) persuasively argue that the science behind genetic engineering is incomplete, that most human diseases and complex traits are not genetically determined but shaped by epigenetic and dynamic processes. According to Richard Strohman, "Most human diseases, and complex traits in all organisms, depend on nongenetic processes. They are shaped by environmentally sensitive regulatory networks of molecular agents that obey dynamic rules. These 'epigenetic' networks are generally unappreciated, and their rules are little understood by modern biotechnology. To prematurely initiate large-scale genetic engineering—whether of vast areas of cropland or of human beings—based on genetic knowledge but epigenetic ignorance is to practice incomplete science and to invite disasters of unknown proportion."
9. Biotechnology industries are also extensions of information industries. As Bill Gates said, "This is the information age, and biological information is probably the most interesting information we are deciphering and trying to decide to change. It's all a question of how, not if." In Silicon Valley, the leading edge is not in building computers but in telling computers what to do—the who's who of new money are software designers. Three young entrepreneurs—

Krishnamurthy, Piturro, and Kissel—are working to create digital clones with "reality merge" functions that will enable people to translate their bodies into digital data to go shopping and have clothes fitted on-line, seek medical advice, or play interactive games. Such a view of information and of the body's information as linked to providing newer, more efficient and powerful "options" for the modern person feeds into the globalizing economy and electronic herd described by Thomas Friedman in *The Lexus and the Olive Tree.* Ideologically, biotechnology as an information technology ignores a natural world in which animate life is part of larger-than-self, nonarbitrary, non–socially constructed, nonvirtual processes.

10. Biotechnology is, as USDA secretary Dan Glickman said, "the Battle Royale" for the twenty-first century. Beyond its battle for markets and control, such a battle fundamentally exposes rapidly diverging worldviews between those often referred to as "globalists" (a.k.a. *Homo economicus*) and those often referred to as "localists" (a.k.a. *Homo eroticus*). Globalists tend to view the living world in economic terms while localists tend to view economics in terms of the living world. As biotechnology is powerfully convergent with a globalist worldview and stands to create entirely new terms and conditions for the living world, the burden of articulation and argument falls on localists. This is a world historical debate that forces to the surface heretofore taken-for-granted or dimly intuited meanings and understandings of the living world. Its kind and quality will depend on all the arts of civilization—science, literature, philosophy, theology, history—and the public's demand for it.

PIED BEAUTY

Gerard Manley Hopkins

Glory be to God for dapples things—
For skies of couple-colour as a brindled cow;
For rose-moles all in stipple upon trout that swim;
Fresh-firecoal chestnut-falls; finches wings;
Landscape plotted and pieced—fold, fallow, and plough;
And all trades, their gear and tackle and trim.

All things, counter, original, spare, strange;
Whatever is fickle, freckled (who knows how?)
With swift, slow; sweet, sour; adazzle, dim;
He father's-forth whose beauty is past change:
Praise him.

WENDELL BERRY

THY LIFE'S A MIRACLE

The expressed dissatisfaction of some scientists with the oversimplifications of commercialized science has encouraged me to hope that this dissatisfaction will run its full course. These scientists, I hope, will not stop with some attempt at a merely theoretical or technical "correction," but will press on toward a new, or renewed, propriety in the study and the use of the living world.

No such change is foreseeable in the terms of the presently dominant mechanical explanations of things. Such a change is imaginable only if we are willing to risk an unfashionable recourse to our cultural tradition. Human hope may always have resided in our ability, in time of need, to return to our cultural landmarks and reorient ourselves.

One of the principle landmarks of the course of my own life is Shakespeare's tragedy of *King Lear.* Over the last forty-five years I have returned to *King Lear* many times. Among the effects of that play—on me and, I think, on anybody who reads it closely—is the recognition that in all our attempts to renew or correct ourselves, to shake off despair and have hope, our starting place is always and only our experience. We can begin (and we must always be beginning) only where our history has so far brought us, with what we have done.

Lately, my thoughts about the inevitably commercial genetic manipulations already in effect or contemplated have sent me back to *King Lear* again. The whole play is about kindness, both in the usual sense and in the sense of truth-to-kind, naturalness, or knowing the limits of our specifically *human* nature. But this issue is dealt with most explicitly in an episode of the subplot, in which the Earl of

Gloucester is recalled from despair so that he may die in his full humanity.

The old earl has been blinded in retribution for his loyalty to the king, and in this fate he sees a kind of justice, for as he says, "I stumbled when I saw." He, like Lear, is guilty of hubris or presumption, of treating life as knowable, predictable, and within his control. He has falsely accused and driven away his loyal son Edgar. Exiled and under sentence of death, Edgar has disguised himself as a madman and beggar. He becomes, in that role, the guide of his blinded father, who asks to be led to Dover where he intends to kill himself by leaping off a cliff. Edgar's task is to save his father from despair, and he succeeds, for Gloucester dies at last " 'Twixt two extremes of passion, joy and grief. . . ." He dies, that is, within the proper bounds of the human estate. Edgar does not want his father to give up on life. To give up on life is to pass beyond the possibility of change or redemption. And so he does not lead his father to the cliff's verge, but only tells him he has done so. Gloucester renounces the world, blesses his supposedly absent son Edgar, and, according to the stage direction, "Falls forward and swoons."

When Gloucester returns to consciousness, Edgar speaks to him in the guise of a passerby at the bottom of the cliff, from which he pretends to have seen Gloucester fall. Here he assumes explicitly the role of spiritual guide to his father.

Gloucester, dismayed to find himself still alive, attempts to refuse help: "Away, and let me die."

And then Edgar, after an interval of several lines in which he represents himself as a stranger, speaks the filial (and fatherly) line about which my thoughts have gathered:

"Thy life's a miracle. Speak yet again."

This is the line that calls Gloucester back—out of hubris, and the damage and despair that invariably follow—into the human life of grief and joy, where change and redemption are possible.

The power of that line read in the welter of innovation and speculation of the bioengineers will no doubt be obvious. One immediately recognizes that suicide is not the only way to give up on life. We know that creatures and kinds of creatures can be killed, deliberately or inadvertently. And most farmers know that any creature that is sold has in a sense been given up on; there is a big difference between selling this year's lamb crop, which is, as such, all that it can be, and selling the breeding flock or the farm, which hold the immanence of a limitless promise.

A little harder to compass is the danger that we can give up on life also by presuming to "understand" it—that is, by reducing it to the terms of our understanding and by treating it as predictable or mechanical. The most radical influence of reductive science has been the virtually universal adoption of the idea that the world, its creatures, and all the parts of its creatures are machines,—that there is no difference between creature and artifice, birth and manufacture, thought and computation. Our language, wherever it is used, is now almost invariably conditioned by the assumption that fleshly bodies are machines full of mechanisms, fully compatible with the mechanisms of medicine, industry, and commerce, and that minds are computers fully compatible with electronic technology.

This assumption may have begun as metaphor, but in the language as it is used (and as it affects industrial practice) it has evolved from metaphor through equation to identity. And this usage institutionalizes the human wish, or the sin of wishing, that life might be, or might be made to be, predictable.

I have read of Werner Heisenberg's principle that "Whenever one treats living organisms as physiochemical systems, they must necessarily behave as such." I am not competent to have an opinion about the truth of that. I do feel able to say that whenever one treats living organisms as machines, they must necessarily be perceived to behave as such. And I can see that the proposition is reversible: Whenever one perceives living organisms as machines, they must necessarily be treated as such. William Blake made the same point very early in this age of reduction and affliction:

> *What seems to Be, Is, To those to whom*
> *It seems to Be, and is productive of the most dreadful*
> *Consequences to those to whom it seem to Be . . .*
>
> (Blake, *Complete Writings*, Oxford, 1966, p.663)

For quite a while it has been possible for a free and thoughtful person to see that to treat life as mechanical or predictable or understandable is to reduce it. Now, almost suddenly, it is becoming clear that to reduce life to the scope of our understanding (whatever "model" we use) is inevitably to enslave it, make property of it, and put it up for sale.

This is to give up on life, to carry it beyond change and redemption, and to increase the proximity of despair.

Cloning—to use the most obvious example—is not a way to improve sheep. On the contrary, it is a way to stall the sheep's lineage and make it unimprovable. No true breeder could consent to it, for true breeders always have their farm and their market in mind, and are always trying to breed a better sheep. Cloning, besides being a new method of sheep-stealing, is only a pathetic attempt to make sheep predictable. But this is an affront to reality. As any shepherd would know, the scientist who thinks he has made sheep predictable has only made himself eligible to be outsmarted.

The same sort of limitation and depreciation is involved in the proposed cloning of fetuses for body parts, and in other extreme measures for prolonging individual lives. No individual life is an end in itself. One can live fully only by participating fully in the succession of the generations, in death as well as in life. Some would say (and I am one of them) that we can live fully only by making ourselves as answerable to the claims of eternity as to those of time.

The problem, as it appears to me, is that we are using the wrong language. The language we use to speak of the world and its creatures, including ourselves, has gained a certain analytical power (along with a lot of expertish pomp) but has lost the power to designate what is being analyzed or to convey any respect or care or affection or devotion toward it. As a result we have a lot of genuinely concerned people calling upon us to "save" a world which their language simultaneously reduces to an assemblage of perfectly featureless and dispirited "ecosystems," "organisms," "environments," "mechanisms," and the like. It is impossible to prefigure the salvation of the world in the same language by which the world has been reduced and defaced.

By almost any standard, it seems to me, the reclassification of the world from creature to machine must involve at least a perilous reduction of moral complexity. So must the shift in our attitude toward the creation from reverence to understanding. So must the shift in our perceived relationship to nature from that of steward to that of absolute owner, manager, and engineer. So even must our permutation of "holy" to "holistic," the latter term implying not mystery but understandability in the relation of part to whole.

At this point I can only declare myself. I think that the poet and scholar Kathleen Raine was correct in reminding us that life, like holiness, can be known only by being experienced. To experience it is not to "figure it out" or even to understand it, but to suffer it and rejoice in it as it is. In suffering it and rejoicing in it as it is, we know that we do not and cannot understand it completely. We know, moreover, that we do not wish to have it appropriated by some individual or group's claim to have understood it. Though we have life, it is beyond us. We do not know how we have it, or why. We do not know what is going to happen to it, or to us. It is not predictable; though we can destroy it, we cannot make it. It cannot, except by reduction and the grave risk of damage, be controlled. It is, as Blake said, holy. To think otherwise is to enslave life and to make not humanity but a few humans its predictably inept masters.

We need a new Emancipation Proclamation, not for a specific race or species but for life itself—and that, I believe, is precisely what Edgar urges upon his once presumptuous and now desperate father:

Thy life's a miracle. Speak yet again.

Gloucester's attempted suicide is really an attempt to recover control over his life—a control he believes (mistakenly) that he once had and has lost:

O you mighty gods!
This world I do renounce, and in your sights
Shake patiently my great affliction off.

The nature of his despair is delineated in his belief that he can control his life by killing himself, which is a paradox we will meet again three and a half centuries later at the extremity of industrial warfare, when we believed that we could "save" by means of destruction.

Later, under the guidance of his son, Gloucester prays a prayer that is exactly opposite to his previous one—

You ever-gentle gods, take my breath from me,
Let not my worser spirit tempt me again
To die before you please!

—in which he renounces control over his life. He has given up his life as an understood possession and has taken it back as miracle and mystery. And his reclamation as a human being is acknowledged in Edgar's response: "Well pray you, father."

It seems clear that humans cannot significantly reduce or mitigate the dangers inherent in their use of life by accumulating more information or better theories or by achieving greater predictability or more caution in their scientific and industrial work. To treat life as less than a miracle is to give up on it.

I am aware how brash this commentary will probably seem, coming from me, who has no competence or learning in science. The issue I am attempting to deal with, however, is not knowledge but ignorance. In ignorance I believe I may pronounce myself a fair expert.

One of our problems is that we humans cannot live without acting; we *have* to act. Moreover, we *have* to act on the basis of what we know, and what we know is incomplete. What we have come to know so far is demonstrably incomplete, since we keep on learning more, and there seems little reason to think that our knowledge will become significantly more complete. The mystery surrounding our life probably is not significantly reducible. And so the question of how to act in ignorance is paramount.

Our history enables us to suppose that it may be all right to act on the basis of incomplete knowledge *if* our culture has an effective way of telling us that our knowledge is incomplete, and also of telling us how to act in our state of ignorance. We may go so far as to say that it is all right to act on the basis of sure knowledge, since our studies and our experience have given us knowledge that seems to be pretty sure. But apparently it is dangerous to act on the assumption that sure knowledge is complete knowledge—or on the assumption that our knowledge will increase fast enough to outrace the bad consequences of the arrogant use of incomplete knowledge. To trust "progress" or our putative "genius" to solve all the problems that we cause is worse than bad science; it is bad religion.

A second human problem is that evil exists and is an ever-present and lively possibility. We know that malevolence is always ready to appropriate the means that we have intended for good. For example, the technical means that have industrialized agriculture, making it (by very limited standards) more efficient and productive and easy, have also made it more toxic, more violent, and more vulnerable—have made it, in fact, far less dependable if not less predictable than it used to be.

One kind of evil certainly is the willingness to destroy what we cannot make—life, for instance—and we have greatly enlarged our means of doing that. And what are we to do? Must we let evil and our implication in it drive us to despair?

The present course of reductive science—as when we allow agriculture to be invaded by the technology of war and the economics of industrialism—*is* driving us to despair, as witness the incidence of suicide among farmers.

If we lack the cultural means to keep incomplete knowledge from becoming the basis of arrogant and dangerous behavior, then the intellectual disciplines themselves become dangerous. What is the point of the further study of nature if that leads to the further destruction of nature? To study the "purpose" of the organ within the organism or of the organism within the ecosystem is *still* reductive if we do so with the assumption that we will or can finally figure it out. This simply captures the world as the subject of present or future "understanding," which will become the basis of further industrial and commercial optimism, which will become the basis of further exploitation and destruction of communities, ecosystems, and local cultures.

I am not of course proposing an end to science and other intellectual disciplines, but rather a change of standards and goals. The standards of our behavior must be derived not from the capability of technology but from the nature of places and communities. We must shift the priority from production to local adaptation, from power to elegance, from costliness to thrift. We must learn to think about propriety in scale and design as determined by human and ecological health. By such changes we might again make our work an answer to despair.

CHRIS DESSER

UNNATURAL SELECTION OR BAD CHOICE

The arrogance of human action in the face of human ignorance is unfailingly breathtaking, particularly so in the brave new world of biotechnology. Upping the ecological ante from the lessons of petrochemicals and their unexamined and unintended consequences (dependence on fossil fuels, ozone depletion, and global warming), the specter of biotechnology is even more dramatic—it gives us the ability to write and rewrite life itself and its evolutionary processes. Today, with little to no public debate in the United States, biotechnology supports a vast industry and is widely in use. It is being deployed to grow crops, alter food, manufacture biological and genetically targeted weapons, create fuel, and develop medicine and medical procedures. Unchecked, it will leave no aspect of our world or our lives untouched. We must urgently ask: How should we begin a meaningful public debate on biotechnology? How shall we appropriately assess its benefits and consequences? How do we weigh the decisions we make and the actions we take in the light of our knowledge and in the face of our ignorance?

My position is not that biotech shouldn't exist. We are not going to stuff the genie back in the bottle. Nor do I subscribe to inherently dualistic ethical, moral, or religious views that humans should not tamper with God's creation. We are inextricably woven into the world and with every interaction our world is changed, as are we. Nor do I have any interest in squelching human creativity or ingenuity or making realms of knowledge off-limits. In fact, I am interested in opportunities for knowledge to keep on evolving, especially scientific knowledge as counsel to the crises we face. However, we cannot ignore the irony that biotechnology actually threatens opportunities for knowledge and creativity. Known as the "Life Industry," biotechnology, by its very nature, controls and diminishes the richness, diversity, and inherent creativity of processes larger than ourselves and our commerce.

The economic success of the "Gene Giants," Monsanto, Novartis, Du Pont, and others, depends upon control of nature and people. The Gene Giants acquire control over people in many ways: by requiring farmers to sign licensing agreements to grow crops, by depriving people of the right to know what they are eating, and by patenting life for their exclusive use and gain. In short, the Gene Giants seek to control life itself by arrogating to themselves the process of evolution. Others in these pages will examine the troubling social impacts of biotechnology: the implications of a few corporations controlling the world's food; the health consequences attendant to consuming genetically modified food; the violation of consumer rights to know the content of the food we eat; the inequities resulting from biopatenting; and the morality of the commodification of life. My interest here is the disruptive impact of biotechnology, especially transgenics, on the ecosystem and the continued unfolding of evolution.

Transgenics, or recombinant DNA technology (R-DNA), is the process that transfers genes between organisms that would not naturally interbreed, creating "genetically modified organisms." For example:

• To preserve strawberries for storage and during transportation over long distances, scientists have implanted flounder genes into the fruit, reasoning that the gene that allows flounders to survive in icy cold water would confer the same benefit on strawberries.

• Hamster genes have been inserted in tobacco for the purpose of increasing sterol production.

• Spider genes have been inserted into goats to enable goats to produce silk in their milk for the production of "Biosteel," a hardy fabric apparently much in demand in the fields of aerospace, engineering, and medicine.

• Human sperm–producing genes have been inserted into rats. The rats then generated human sperm. Scientists are now seeking permission to use this rat-man sperm to fertilize human eggs. This apparently has something to do with enhancing male fertility.

Transgenics thus transfers an "isolated" genetic quality from one species to another for an alleged benefit. To date, however, there has been little research and less understanding of the ecosystemic effects of such genetic manipulation, or of the meaning and significance of a gene in context.

The original white paper establishing U.S. biotech policy is remarkably unconcerned with questions addressing the ecosystemic impacts of genetic modification ("Introduction of Recombinant DNA–Engineered Organisms into the Environment: Key Issues, Prepared for the Council of the National Academy of Sciences Committee on the Introduction of Genetically Engineered Organisms into the Environment," National Academy Press, Washington, D.C., 1987). It concludes that "there is adequate knowledge of the relevant scientific principles, as well as sufficient experience with R-DNA-engineered organisms, to guide the safe and prudent use of such organisms outside research laboratories." Unbelievably, it found that "there is no evidence that unique hazards exist either in the use of R-DNA techniques or in the transfer of genes between unrelated organisms," and that "the risks associated with the introduction of R-DNA-engineered organisms are the same in kind as those associated with the introduction into the environment of unmodified organisms and organisms modified by other genetic techniques." (Small solace. The hazards of the introductions of what we now call weeds or exotics or nonindigenous species are well documented, including the unanticipated havoc wrought by zebra mussels, kudzu, water hyacinths, and other species in various ecosystems throughout the United States.)

The White Paper further asserts that "mounting concerns about environmental degradation, together with the pressing problems of ensuring adequate food and health care for a rapidly expanding global population, provide a compelling rationale for the accelerated study and development of biological organisms for use in agriculture, health care, and biosphere management. The committee concludes that R-DNA techniques constitute a powerful and safe new means for the modification of organisms." Ironically, it is more likely that R-DNA technologies will exacerbate the very maladies the committee believes they will cure. I am particularly concerned here with ecological degradation.

The rationale supporting the development and application of transgenic technologies is embedded in a linear and instrumental view of evolution and biology. Rather than seeing a gene as part of a complex living system, both microcosmically and macrocosmically, the transgenics complex values a gene only for a narrow and singular function. According to Russel Howard, president of Maxygen, a leader in the field of "directed evolution," the attitude in biotechnology is "here are genes from nature, what can I do to squash these into a workable commercial product?" Systems thinking, which entails an understanding of ecological processes, is quickly becoming a keystone of our contemporary understanding of biology. Here, it is utterly absent.

On the micro level, transgenic technologies disregard the many ways in which a given gene may function within an organism. Fritjof Capra stated in *The Web of Life* that "research has shown . . . that a single gene may affect a wide range of traits and that, conversely, many separate genes often combine to produce a single trait. . . . The study of the coordinating and integrating activities of the whole genome is of paramount importance, but this has been hampered severely by the mechanistic outlook of conventional biology. Only recently have biologists begun to understand the genome of an organism as a highly interwoven network and to study its activities from a systemic perspective."

In the strawberry example above, a systemic perspective would examine the effect of the new gene on the plant over time. How will the flounder gene change the path of the development of the strawberry as an integral structure, as a system? In a recent *New York Times Sunday Magazine* article, Michael Pollan raised the issue this way: "The introduction into a plant of genes transported not only across species but whole phyla means that the wall of that plant's essential identity—its irreducible wildness, you might say—has been breached." Surely this has consequences.

On the macro level, transgenics rips the gene out of species/context A and then violates the integrity of the context, or ecosystem, and structure of to-be-altered species B. It ignores all the environmental interactions that, over millennia, have given rise to the unique qualities that characterize species A. Further, it ignores the impact that the genetically modified organism will have on all the processes and organisms forming the ecosystem within which it exists. To return to the strawberry example: How will the ecosystem within which the strawberry is grown be altered? Will the strawberry grow out of control and become a weed, displacing other plants? How will strawberry pollinators and the other plants they come into contact with be affected? How will animals, including humans, who eat the strawberry be affected?

Physicist Peter R. Willis sums the issue up nicely: "The meaning of a gene is determined by the context in which it is expressed. It also contributes to that context." The Gene Giants forge ahead with a terribly partial understanding of genetics, evolutionary biology, and ecology. They remain unconcerned with the meaning of a gene except in the narrowest and most mechanistic sense, and there are ramifications. The destructive impact of genetically modified organisms on all levels of ecosystems, from water to soil to insects to birds, is being increasingly documented. Here are a few examples:

- In 1994, soil researchers Elaine Ingham and Michael Holmes formulated a new system to test a genetically engineered bacterium designed to help turn crop wastes into ethanol. They found that the genetically modified organism also had a devastating impact on crops. An unanticipated consequence of the application of the bacterium was the destruction of the mycorrhizal fungus essential to plant cycling of nitrogen. According to Ingham, if the bacterium had been released and spread widely, growing crops without a control measure for the organism would likely have become impossible. Tests did not provide adequate information for evaluating the impact of these biotech products in complex, real-world systems like soils.

• More recent ecosystemic impacts have been documented in Great Britain. In March 1999, *The Guardian* reported new concerns about genetically modified crops when scientists found that, not surprisingly, the life span and fertility of ladybugs was significantly shortened when their food chain, specifically aphids, was poisoned by genetically modified crops. The scientists fed genetically modified potato plants to aphids, which were then fed to ladybugs. The ladybugs' lives were shortened by up to half, and their fertility and egg-laying was significantly reduced. Females were apparently affected more seriously than males, and a change of diet to aphids not exposed to genetically modified plants seemed to reverse the process. The scientists concluded that "altering the genetic make-up of plants to resist destructive aphids might have serious effects on other, natural pest-killers [prompting] new questions about the food chain for birds. None is known to eat [ladybugs], but several, including bluetits and warblers, feed on aphids. [Ladybugs] are traditionally regarded as gardeners' and farmers' best friends and their well-being is a prime indicator of environmental stability."

• *Bacillus thuringiensis* (Bt) is a natural bacterium that is toxic to some insects. It has been used by organic farmers to control crop-damaging insects. In recent years it has been genetically manipulated into various plants, including maize, in an effort to reduce the need for pesticides. But once again, there were unexpected consequences. Laboratory tests conducted by Angela Hilbeck, from the Swiss Federal Research Station for Agroecology and Agriculture, found that beneficial insects were also killed by Bt—lacewings died after eating corn-borer caterpillars that had consumed the toxin. In a Kafkaesque response, Novartis, the company that produces Bt maize, countered that they had conducted extensive experiments and that Hilbeck's lab did not reflect "real-world" conditions (which, if anything, are more complex). Hilbeck explained Novartis's error: In standard tests, "an insect is fed eggs sprayed with Bt-toxins. The problem is, however, that lacewings do not eat the eggs but suck them out from inside, so they do not [come into contact] with the poison on the outside of the egg. Only with long-time feeding trials and a very careful set-up of the experiment can such impacts be studied. And such experiments have not been done."

Emulsions from Bt-bacteria have been safely used as sprayable insecticides since the 1950s. But there are significant differences between sprayable Bt and Bt-modified crops, as explained in a World Watch paper (January/February 1999): "[W]hile the naturally derived spray version of Bt is highly specific (its toxicity is activated only in the gut of certain species), the genetically modified version has been altered to work against an array of insects—harmful or not." World Watch also quotes studies indicating that beneficial insects and soil organisms were also harmed by Bt and that "preliminary studies suggest that the adverse effects could even be felt by insect-eating bird populations, many levels up the farm foodweb—a foodweb that includes plants and animals consumed by humans."

The manner of application presented a second significant difference. According to the same World Watch paper, the Bt crops delivered much higher levels of the toxin continuously, "roughly 10 to 20 times the lethal dose of sprayable formulations, in contrast to the carefully timed and dosed applications of sprayable Bt. . . . These transgenic crops are now pumping out huge amounts of toxins from all tissues throughout the entire growing season, from germination to senescence."

This case demonstrates why many factors must be considered before an accurate analysis of the ecosystemic effect of genetically modified crops can begin to be understood. Among the variables that Novartis ignored were the means of insect ingestion; unintended insect, bird, and animal victims; the form and dosage of the toxin; and how the timing of the release of the poison might be affected by season and plant cycle.

For monarch butterflies it may be too late. In a Cornell University study in May 1999, researchers found that Bt pollen was deadly to monarch butterfly caterpillars. Entomologist John E. Losey and his research team fed monarch caterpillars milkweed dusted with Bt corn pollen, non-Bt pollen, and no pollen. Half the caterpillars fed the Bt-dusted pollen died within four days of consumption; none of the other caterpillars died. In a *New York Times* article (May 20, 1999), Dr. Fred Gould, insect ecologist at North Carolina State University, said, "Nobody had considered this before. Should we be concerned? Yes."

Obviously, these issues are no longer theoretical or limited to laboratory experiments. In February 1999, Monsanto, one of the world's largest engineers of genetically modified foods, was fined £17,000 (around $27,540) by British authorities for failing to stop an altered crop from escaping into the environment (a paltry sum considering the magnitude of the breach of trust and the potential ecological consequences). This season ten to twenty million acres of Bt corn have been planted in the corn belt, the heart of the monarch's breeding range. This summer, here in the United States, we will witness the consequences for the monarch and other wild insects.

The British press, unlike its American counterpart, has covered biotechnology extensively. Mounting evidence about the environmental consequences of genetically modified organisms and strong public pressure is causing the British government to adopt a more cautious view than that of the United States. In March 1999, *The Guardian* wrote about a British government report that identified potentially adverse effects of releasing genetically modified crops into the environment, stating: "The persistence, invasiveness and competitiveness of new species could change the population dynamics of surrounding areas by overwhelming native plants and reducing the animal species that depend on them for survival."

Wind or insects could transfer inserted genetic material to native plants, turning them into hybrids with selective advantages over other native plants, which may then suffer. Soil decomposition may be affected by changed nitrogen and carbon recycling processes.

The law of unintended effect may result if genetically modified plants unexpectedly turn out to be unpalatable to herbivores, a trait that could be transferred to native species.

The Gene Giants create designer species for particular commercial purposes. I have mentioned a few above—resistance to predators and greater tolerance for heat or cold—but there are many others: faster growth, bigger or smaller size, better shape for shipping, more pleasing taste or color, to name a few. It is entirely foreseeable that any genetic change that might find a bigger, better market can probably be achieved. Those species are patented and then become private corporate property and replicable economic commodities. As we know, successful large-scale replication requires predictability, uniformity, and standardization. And economic benefit requires exclusive ownership and control over the manufacture, use, and distribution of the commodity. Without public debate and authoritative overview, biotechnology represents the worst combination of the mad scientist and the unrestrained capitalist.

Further, economic gain for the few who "own" the life they are manipulating has grave consequences for humans and other species. As the British report recognizes, the biodiversity characteristic of healthy ecosystems, and the genetic complexity that enhances evolutionary options, will be reduced if the many threats that transgenic crops pose to insects, animals, and other plants, both wild and cultivated, come to pass. Diversity will be further diminished as farmers around the world give up growing indigenous plants in favor of "modern" crops like Bt maize.

The Gene Giants, as we know them, disregard the impact of their actions on the web of relationships that constitute our global ecosystem when, in fact, neither they nor we have any idea how altering a species will affect complex ecosystems or evolution. How will these new species evolve—the designers concern themselves only with the first generation of these plants, but how will subsequent generations be affected? Having changed the genetic structure of an organism, how has the course of that organism's evolution been altered? How will the ecosystem be changed? In the face of these questions, and despite this uncertainty, they ignore the Precautionary Principle, which holds that when an activity raises threats of harm to human health or the environment, precautionary measures should be taken even if some cause and effect relationships are not fully established. Amory Lovins, co-founder of Rocky Mountain Institute and co-author of *Natural Capitalism,* recently observed (June 1999) that "ignoring or removing the 90+% of the genome whose function is unknown is like monocultural agriculture that herbicides out the biodiversity. That's the end of proper ecosystem functioning and of resilience. The ignorant may get away with it briefly, but then nature catches up with them." And "them" is not only the Gene Giants but all of us.

Steve Jones, professor of genetics at University College in London, believes that genetic pollution, or "drift," is inevitable, that genes from genetically manipulated plants will mix with other plants with potentially devastating consequences. "Everybody knows that," says Professor Jones, "and we have no idea what is going to happen." Should the gene be one that confers resistance to insects, "suddenly we have no insects. With no insects you have no ecology, no ecosystem, no pollinators, no flowers, God knows what."

Professor Jones continued, "Now this probably will not happen. But it certainly might. With genetically modified plants, we are doing something new. We are moving genes around to where they've never been before, and we don't seem to be considering the possibility that evolution may take advantage of those genes, as it has done previously. In ways that we don't like."

Most of what we have learned as a species, including the very means to develop biotechnology, has come from our observation of the natural world, which continues to teach us. Ironically, the more we know, the more we come to see how little we know. Through such observation over thousands of years, we have learned that nature functions as a system, and that system is violable. Biotechnology not only poses the greatest violation yet to our ecosystem, it also threatens the integrity of our greatest text and teacher. The consequences of biotechnology going awry are grave and irrevocable—humans, like every other living thing on the planet, will be affected in as yet unpredictable ways. In one generation and after millions of years of evolution, we may well ruin our greatest source of knowledge.

STUART NEWMAN

(Interview)

EPIGENETICS VS. GENETIC DETERMINISM

Casey Walker: As a cellular biologist, where do you see poor assumptions and bad theory playing themselves out in biogenetic engineering?

Stuart Newman: It begins with the false idea that organisms can be designed to specification, or corrected by popping out bad genes and popping in new genes. We see these assumptions in agriculture with genetically engineered foods, and with practices such as inserting naturally occurring insecticide proteins into crop plants like corn. There is an incorrect, but unfortunately prevalent notion, that genes are modular entities with a one-to-one correspondence between a function and a gene. My particular interest is in how these ideas are being played out in human biology, where we see the same kind of genetic reductionism justifying attempts to assign genes to complex conditions such as schizophrenia, intelligence, homosexuality, and so forth. Definition of problems in genetic terms obviously leads to calls for genetic solutions with profound consequences for human beings and evolution.

Although it's unquestionable that every complex biological condition has a genetic component to it, the mediation that occurs between the genetic component and the actual behavior or feature is typically quite complex and should militate against taking a reductionist approach. Frequently, a gene in one context will influence a condition in one way, and in a different context will influence the condition in a completely different way. There's simply very bad theory behind a lot of the genetic interventions now being proposed. In particular, bad theory (tied to commercial interest) is at the root of proposals for human germline modification, which would take a human embryo on the path to developing one condition or another, perhaps a disease, and modify its genes.

Is it misleading to perceive genetic expression and environmental influence as two discrete processes?

Yes. There's a genetic component to an organism's susceptibility to environmental effects, and there's an environmental component to its expression of genetic effects. Thus, there's a composite of interpenetrating genetic and environmental processes that give rise to every organism during development. Another very common misperception comes with the conclusion that anything congenital—inborn—is inherited from the parents' genes. There are many studies currently attempting to tie personality traits such as shyness or aggression to genes. While people often recognize that various traits seen in their children were there from the very start, they need also to recognize and understand that the developmental processes of that child were far more complex than a playing out of its inherited genes. There are infinitely complex processes during development that make each outcome unique. Features in a newborn that are undeniably congenital and could even be said to be "hardwired" into the biology of that person may have very little to do with either parent. Thus, to say something is "congenital" does not mean it can be deconstructed and attributed to inheritance from one or the other parent.

Will you describe those processes that influence various outcomes during development of the human embryo?

At the start of development, the fertilized egg has all the nuclear genes contributed by its parents, and also the separate mitochondrial genes from the mother in the cytoplasm of the egg. The egg's cytoplasm also contains protein and RNA products of some of the mother's genes that are not part of her genetic contribution to the embryo. At first, the genes in the nucleus of this newly fertilized egg start to be activated and cause proteins to be made. But that's not the only thing that's going on. There's the mother's uterine environment that this organism is exposed to, and there is an "intrinsic plasticity" that allows the embryo to readjust and recover from perturbations or disturbances. For example, if you have a two-cell embryo and somehow the cells get detached from one another, each of those two cells—even though both were originally one-half of an individual—will go ahead and become a separate, complete individual. This, of course, is the basis of twinning. Mouse embryos at the two-cell stage can be separated and each of the cells will make an individual, even though under natural circumstances they wouldn't have done it. Through this kind of plasticity, a species-characteristic outcome is achieved even if it now takes the form of two organisms.

Something even more unusual can be done experimentally that may never or only rarely happen in natural circumstances, which is to take two embryos that are separate from each other and jumble the cells together. Again, these cells will readjust their fates to create a complete individual. You can show this by taking organisms of two different strains, or two different species, and creating one single organism from them. You can make a chimera, which is what these combined embryos are called, between a sheep and a goat (a "geep"). Of course, that would never happen in nature, yet we do get a composite individual with all the normal parts.

Which implies a kind of strategy or will within embryonic cells?

It's a subject of major, scientific debate as to what it implies. Some say that throughout our evolution embryos and organisms have been subjected to so many different stresses and strains and aberrant environments and strange conditions that we have within us a completely hardwired set of programs to get us out of all of these things that might happen. This notion has been put forward by some developmental biologists recently and called "adaptability." They say this developmental plasticity is a very sophisticated product of our evolutionary history, and is dependent upon highly evolved genetic circuits and programs.

That's not the view I take. I see plasticity, or the ability to readjust in the face of environmental change and to take on characteristic forms despite all the vicissitudes of the developmental process, as a property intrinsic to the materials that make up organisms. An analogy may help here. If you look at rain, you'll see that every raindrop falling through the air has the same shape. Why is that? Not because a raindrop has genes to develop its shape, but because it's a piece of a particular kind of matter, a drop of water being subjected to certain external, physical processes. If you take a still body of water and agitate it, you will always make waves; if you swirl it, you will always make vortices. Here, too, a particular material will do a certain set of stereotypical or "generic" things because of its composition and the forces to which it is susceptible. There are many more sophisticated properties that certain materials can exhibit, even if they're not alive, that support this view. There's a whole class of materials called "excitable media" that are studied by physicists and chemists. These are things that will give you back more than you put into them because they contain stored energy and have a stored ability to react chemically. For example, chemical reactions of diffusing molecules can spontaneously produce stripes or spots or spirals of chemical substances arranged across a spatial domain. Since this occurs with nonliving materials, we know there's something characteristic about excitable material itself that is not simply the result of a list of ingredients (which is what the genes provide). Instead, the composite materials formed from those ingredients will exhibit certain generic physical behaviors.

Now, embryos are excitable media. They inevitably do certain things because of their physical and chemical properties. This opens up a whole new set of causalities in the formation of an organism. It's not simply tracking the playing out of genes but, rather, recognizing that there are physical and chemical properties that arise as the products of genes interact with each other within cellular and multicellular contexts that also contain nongenetic substances—water, ions, and so forth.

Is there a threshold, a critical mass of cells, where an embryo becomes an "excitable medium"?

The excitability is there from the start because each individual cell is excitable. It's metabolizing, it's exchanging matter and energy with its environment. But individual cells have a limit as to how differentiated each can get. Even though a single cell has many substances in it, and these substances may be produced in one part of the cell and not the other, there will be a rapid mixing and homogeneity that will generally prevail because the cell is so small. However, with a cluster of cells, because it is larger, something may occur or be produced at one end and not at the other that creates the gradients or inhomogeneities which provide the basis for cells to differentiate and take on distinct roles. Increased size brings several new physical factors into play that affect cellular development. In addition to the role of diffusion just described, surface tension begins to play a role in embryo shape and tissue boundary formation. This is another factor that is not relevant at the scale of the individual cell.

Now, if it is the case that in embryogenesis of a contemporary organism these physical processes have an importance that is neglected by concentrating solely on genetics, it is also true that evolution has utilized the outcomes of those physical processes. In certain cases, the outcomes were adaptive—they led to organisms that were functionally adequate in certain environments—and those outcomes were consolidated by additional processes, often genetic, that built upon and stabilized them. If a physical process led to an outcome that worked in nature, then, after a great deal of evolution, we find there are ways of achieving that endpoint independent of the originating, physical process. This is important because evolution away from strict reliance on physical processes makes morphogenesis more robust and reliable.

When we look at modern organisms, we can see the imprint of these physical processes along with the genetic processes that support and reinforce them. The process of evolution opportunistically consolidates certain outcomes that may have originally arisen as a result of completely different causes. At the end of a long period of time, you have many parallel processes directed toward the same endpoint. In short, the modern organism into which the embryo develops is a very, very sophisticated structure that makes use of genetic processes, physical processes, and genetic processes that have co-opted the outcomes of physical processes in ways physics alone can't do.

If we look back at the raindrop analogy, we see that the raindrop has a head and a tail because of the physical medium it's falling through and the material it's made of. But you can imagine that if the raindrop had genes as well, and those genes were subject to evolution, you might find that after a half billion years that particular shape wouldn't necessarily depend on the continuous falling through the air in order to be maintained. There might be other ways—genetic ways—of getting that shape to be established. It is also clear from this that structures may arise for physical reasons that are not necessarily "adaptive" and later become consolidated by genetic circuitry. These features, which may be as profoundly part of an organism's identity as body cavities or segments, may ultimately have little or nothing to do with adaptation.

When you write of the conceptual gap occurring today in evolutionary theory, is this it?

Yes. If we go back in evolutionary history, before genetic integration and consolidation took place, the interactions between the material of the organism and the external world were very conditional, very context-dependent. Such interaction-dependent causation is called "epigenetic." With virtual certainty, those processes were very important in early evolution. As evolution progresses, the genes capture some of these outcomes and integrate them into the repertoire of the organism, so that what previously depended upon organism-environment interaction is internalized and part of an intrinsic program in the system.

Yet we commonly assume genes generate rather than support various physical features?

Exactly. Even if a feature becomes genetically prescribed, its origin was most typically in an interaction. We can look at the outcome—the end products of evolutionary processes—and appreciate the ways in which genes latched on to all sorts of things that originated through epigenetic mechanisms. However, if you try to understand the structure of the system by just looking at the genes, you will be terribly confused. Life forms did not arise from incremental pathways of small genetic changes. Instead, genes basically insinuated themselves into processes that genes themselves did not originate—a phenomenon ignored in the neo-Darwinian notion of the incrementally achieved "genetic program." In other words, genetic integration is a post hoc scaffold that stabilizes life forms but is very different from a program. Looking at organisms this way allows us to appreciate the fact that there are aspects of our biology that have been consolidated by genetic evolution even though genetic evolution did not originate those aspects.

This becomes interesting when we look at humans and consider our biological repertoire. Even if certain aspects of our biology are completely settled, as in the case of the general form of the human body, that doesn't mean other aspects are. Our brain's physical form results from relatively programmed morphogenetic processes during our development, and yet its cognitive potential remains subject to interactions throughout our lives. New thoughts are not dependent on remolding the brain's morphology, but rather depend upon connections, many of which are conditional-epigenetic. Our brains are not finished products of evolution. The topology of the neuronal connections in the brain is plastic—fluid in the sense of being readjustable based on context. The ideas and values we hold are based on social interactions and interactions with the outside world.

That's not to say that genetic evolution may not eventually consolidate some of these aspects as well. For example, some species of birds learn who their predators are because their parents will squawk when predators come by and they learn to recognize certain silhouettes as hostile. Other birds have an inborn propensity to react very strongly against certain silhouettes. In some lineages of organisms, certain things result mainly from epigenetic interaction and in other lineages they result mainly from genetically based propensities. Now, if we want to interpret what's going on in a reasonable way, it seems as if interaction plays the originating role, and genetics only captures and consolidates the behaviors under certain evolutionary circumstances.

Does this imply that evolution consolidates a certain taken-for-grantedness, a genetic wiring for survival?

It depends on the social and ecological setting in which any lineage finds itself. It may be that under certain evolutionary circumstances things in the experience of that lineage get consolidated into the genome, but it's important to note that consolidation also leads to rigidification. If certain nonhuman species have a hardwired set of behavioral capabilities, they have thereby lost the plasticity that human brains still retain. I would suggest that cognitive plasticity is really, in some sense, a primitive feature that never got rigidified in the human lineage. Although our bodies have become evolutionarily stereotyped, our cognition has not. We've retained the interactive capacity that is probably at the origin of all cognition and behavior, but we've made use of that "primitive" plasticity to a much deeper extent than other species.

Is higher plasticity true of sentient organisms in general?

Humans and dolphins seem to have retained this much more than other organisms, and there's novelty that comes into play with it as well. If primordial organisms indeed had brains that exhibited a lot of behavioral plasticity, they were also very small brains that weren't capable of very high levels of cognition. But if you simultaneously have a large brain and one that has retained behavioral plasticity, you are in very good shape for interacting with your environment in novel and productive ways. Thus our behavior, thoughts, and imagination all depend on organism-organism and organism-environment interaction.

I also must say that this is an area in which I think the evolutionary psychologists and sociobiologists draw incorrect conclusions. Many look at primate species that are supposedly "lower" than we are or more "primitive" evolutionarily—which I think are incorrect ideas—and say that because certain stereotypical behaviors are found in both primates and human beings, these behaviors must be deeply embedded in our genes. I think this is totally wrong. Many behaviors—aggression, territoriality, sexual roles—may arise from circumstances in particular social settings, and initially depend on those social settings for their perpetuation. They may work in allocating resources in a successful fashion under certain constrained conditions. Those circumstances may pertain to certain human societies in our history, as well as to chimpanzees and baboons and so on, yet it's very reasonably the case that whereas these conditional outcomes may have become genetically integrated, consolidated, and hardwired in certain species—rats, baboons—they may still remain dependent on circumstance for humans. Even if a genetically fixed behavior in an ant or a rat looks like a behavior we see in people, it doesn't mean that it's associated with particular genes in a person. This is a common fallacy and, again, comes from not appreciating the role of epigenetics and plasticity in evolution.

Will you address the concept of the "intensification of uniqueness" as opposed to "open-ended production of difference" as another way of looking at evolution?

The standard view of how organisms have evolved, which is the Darwinian view, assumes a general correspondence between genetic change and phenotypic change. There's a kind of uniformitarianism embedded in Darwinism that says that the general progress or alteration of phenotype is correlated with the general rate of alteration of the genotype. If you take that point of view, organisms are always on their way to becoming something else, and any boundaries between species are incidental. In Darwinism, there's a general propensity to think that species identities are transient, temporary distinctions. They look like natural groupings, but the boundaries are always blurry because there's always the possibility of moving outside that perimeter through successive genetic change.

From which transgenics follows easily?

That's right. Darwinists say the idea that species are discrete, separate entities with species boundaries that are not crossed naturally is a remnant of biblical creationist ideas. The point of view that I'm describing, which is based on epigenetic causality, says that at the time the major differences between organisms arose, they did so on the basis of epigenetic changes—what I've also been calling plasticity or conditional and interactive processes. A given genotype would have exhibited a range of phenotypes, depending on the circumstances. In other words, there was no necessary connection between what the genetic content of an organism is and what the organism looks like. Physical and epigenetic determination may have been so important at these early periods of evolution that if the temperature, salinity, or some other aspect of the environment was changed, you would have gotten a very different-looking organism. Now, if the origin of organismal diversity was in epigenetic processes, and if genetic evolution acted upon those dramatically divergent forms and consolidated them under various conditions of life, then after vast amounts of time you would have organisms that were no longer malleable or interconvertible. They would become walled off from each other by the genetic consolidation that evolution produces.

Over time, then, organisms stop changing into other kinds of organisms; they're becoming more themselves. Their characteristics are becoming more and more integrated and intensified so that at the end of a long period of evolution the boundaries between species have become much sharper. This view basically turns neo-Darwinism on its head in its proposal that phenotypic change precedes the genetic evolution that consolidates it. This is possible because most phenotypic innovation results from epigenetic processes. These processes can be mobilized by either environmental alteration or genetic mutation, but any new character that results will be subject to a more gradual co-optation by subsequent genetic evolution. This implies, contrary to neo-Darwinism, that most genetic change doesn't play an innovating role, it plays an integrating and stabilizing role. If you go back to the earliest history of organisms, I think it's inescapable that there was much less genetic integration, much less resistance to perturbation, and you had organismal forms that were malleable and polymorphic, because phenotypes were more dependent upon circumstance.

Such organisms differed from modern ones—their capacity to undergo phenotypic change in response to altered conditions having been virtually Lamarckian. After time, with genetic consolidation, organisms evolved into the Darwinian entities that populate the contemporary biosphere. However, the high degree of genetic integration means that the period of large-scale evolutionary change is over—Darwinian mechanisms of small phenotypic alterations due to small genetic changes will never result in a new genus, class, or phylum.

Along these lines, I appreciated the analogy, in your chapter "Carnal Boundaries," that the organic possibilities of life are as distinct as the elements found on the periodic table.

Right. The periodic table displays the 110 or so stable "types" (elements) that are possible given the physics of the fundamental particles involved. This is all you can get regardless of how much time elapses. I would suggest that in an analogous fashion, the pertinent physical and other epigenetic processes acting upon aggregates of living cells can only give you a predictable, limited array of body types—the core of the taxonomical chart—regardless of how much additional evolution occurs. Of course, living systems are more massive, complex, and multifaceted than atoms, so you can get more subtypes within the major themes as organismic evolution progresses.

When the Darwinian model is upended, what does this imply for human potential?

It does imply a different way of conceiving of potential. I should say that, in substituting an alternative view for Darwinism, it's important to utilize and appeal to concepts that are as rigorous, or more rigorous, than those identified by genetic determinists. Genetic determinism has claimed the scientific high ground because it deals with the very specific, measurable, quantifiable, tangible entities of genes. I would not want to supplant Darwinism with a paradigm based in the metaphoric or metaphysical. The epigenetic view brings in other causal modalities that are neglected in the standard picture, partly because of the training of biologists. Today, a student can go through a university biology program through the Ph.D. without ever taking physics, and miss out on this whole level of causation. The concept of epigenetic-genetic interplay is scientifically more complete than genetic determinism, and genetic determinism is actually obscurantist because it tries to explain things by genes that are genetically inexplicable.

So, to return to your question about human potential, which takes us into the realm of art and poetry, we must recognize that we're dealing with a human brain that, on a physical level, is a highly interactive system in which multiple causalities are brought to bear. To acknowledge this is to be more, not less, scientific. Evolutionary psychologists aspire to rigor by saying it's all in the genes. But if you consider how the evolution of the body occurred over vast amounts of history, and how the outcomes of epigenetic processes have been genetically coopted and assimilated in certain lineages and not in others, it is clear that neural connectivity in the brain has been subject to the same kind of process. Even though we are partly the product of an immense period of genetic evolution, it does not follow that our thoughts and our ability to imagine are the products of genes. Analyzing the human mind genetically is like trying to interpret *The Divine Comedy* by chemically analyzing the ink with which it's printed.

Based on this view of evolution, species boundaries, and epigenetic influences on development, how do you establish lines for what is appropriate or inappropriate in biogenetic engineering?

While it is true that certain versions of genes are associated with certain disease conditions, this is only part of the story. We know the gene and exact site of mutation in sickle cell disease, for example, but we don't know why this disease is mild in some individuals and fatal in others. Similarly for cystic fibrosis and phenylketonuria, the diseases are far more complex than the designation "monogenic trait" would imply. I would say it's rational to use genetic information in a very conservative way—as a prenatal diagnostic. But the idea of using genetic information as a tool to go back to the embryo and start tinkering with it is not rational at all.

How is the cherished goal of biological perfectibility and the eradication of biological defects through genetic engineering a misguided goal for humans and for evolution?

First, it's easy to fall into language that looks at all deviation from certain norms as being a "disease" condition. For example, it's been noted that if everyone were genetically engineered to be six inches taller, there would still be the same number of people in the lowest quartile of height. I'm on the board for the Council for Responsible Genetics, in Cambridge, Massachusetts, which has been considering the social implications of genetic technologies for the past two decades. Although we started with the perspective that the use of genetic information should be left to individuals, we have grown to appreciate how deeply individual thinking about biological variability is influenced by the prevailing eugenic ideology. Even the concept of a birth defect is a relative concept.

It's also pretty clear that our germline and somatic cell genes are under assault by environmental pollutants and the thinning of the ozone layer, and that some birth anomalies are tied to environmental toxins and prescribed and over-the-counter medications. Particular cases have often been difficult to establish because it is frequently impossible to distinguish statistically real effects of known toxins from clusters of cases that are randomly occurring, or due to unknown agents. Furthermore, our knowledge of the basis of vulnerabilities to toxins, or synergistic effects among them, is quite primitive.

In any case, because of the interplay of epigenetics and genetics, it may be impossible, even in principle, to determine if an abnormal developmental outcome was "environmental" or "genetic" in certain cases. (Many cases, of course, will be less ambiguous.) It should be recognized, however, that even if epidemiology does not disclose a clear-cut relationship between a chemical and a type of defect, that does not mean that the chemical did not contribute to the defect. Polluters and manufacturers of suspect drugs will typically want to blame the victim—saying "bad genes" were the cause of an individual's birth anomaly. Since genetic background influences susceptibility to toxic substances, the logical consequence of genetic determinism will be to screen people's genes and tell them where they can work or live, rather than clean up the environment. At a meeting I attended, a well-known Human Genome Project program director actually went so far as to propose inserting extra genes into human embryos in order to give the resulting individuals enhanced capacity to repair environmental damage to their DNA. One can imagine a genetic fix for anything, including the destruction of the natural world. And if it involves doing experiments on human embryos, thereby risking malformations, brain damage, and new forms of cancer, so be it.

Can we say that the set of problems addressed by genetic engineering is not well-posed, that the causes of human suffering have cultural and societal rather than genetic sources?

Exactly. Right. People being outside of the norm one way or another is not the problem. My colleague Gregor Wolbring, a biochemist at the University of Calgary, was a victim of prenatal thalidomide exposure, but he does not consider himself to be a person with defects. There's an interesting connection here with the evolutionary ideas I was discussing before. The way that Darwinism accounts for structural innovation is by the accumulation of many incremental changes that are tested at each step for functional advantage or improved adaptation. But people with congenitally missing limbs and other birth anomalies typically reject prostheses and find a way of operating in the world that suits their biology. When we see how thalidomide people relate to the world, and other people with so-called birth defects, we see they typically find a way of operating that suits them. Organisms don't relate to their environment because they've evolved to match that environment more and more perfectly, but because they figure out how to make what they have work. For example, there was a community on Martha's Vineyard in Massachusetts in which almost everyone was deaf. Deafness was not considered a defect because that's the way the people there were. So, with epigenesis and creative survival—followed, in many cases, by genetic consolidation—driving evolution, we can throw out the incrementalist perfectionism of Darwinism; it's not needed.

Will you comment on the distinctions to be made between a wild system and one that is biogenetically engineered? At what point does engineering usher in irreversible artifice or domestication for a species and a system?

The idea of the "natural" and the "wild" is out of fashion with many geneticists and evolutionary biologists who see evolution as pure opportunism, lacking any inherent direction. Short of inducing an overt pathology, genetically engineering an organism in this view yields a product with no ontological distinction from a naturally occurring organism. While I resist romanticizing the "wild," if evolution proceeds in preferred directions, as I have suggested, it becomes harder to sustain the notion that arbitrary genetic changes are as natural as evolved ones. Domestication of animals has been pointed to as an example of human-guided deviation from the wild. I suggest, however, that like natural evolution, but unlike the results of genetic engineering, the phenotypic changes induced by domestication have proceeded in "natural" directions.

There have been some studies going on for over a half-century in Siberia on the domestication of foxes. The geneticist Dmitri Belyaev, who started the whole enterprise, looked at different domesticated animals and saw commonalities among even widely divergent species that had been domesticated, such as dogs, cattle, and pigs. There was a common reshaping of the skull, and even in the pattern of coloration there was a convergence to certain recurrent themes. Belyaev and his colleagues decided to try it with foxes, a species that had never been domesticated before. They found that in just a couple of generations the same changes occurred that had occurred in other, unrelated lineages. They found that if animals are selected for docility—a common mode of domestication—then the maternal environment of the embryo contains decreased levels of aggression-associated hormones. This, in turn, affects the course of the embryo's development, delaying certain processes and accelerating others, altering fetal form and physiology. While the investigators have hypothesized that their initial selection for docility was a selection for genetic variants, this is just a speculation, although one that is understandable from Russian biologists eager to avoid a Lysenkoist taint. It is also known, however, that phenotypic differences may even exist between genetically identical individuals. What is clear is that the motive force of the morphological changes observed in this study was epigenetic—a changed gestational environment. Moreover, common epigenetic processes seem to be involved in the convergent effects of domestication of genetically divergent species. It is not too much of a stretch to imagine that the transition from ape to human occurred through such epigenetic causation, brought about by self-domestication. After all, we share more than 98 percent of our genetic sequences with chimpanzees. The standard idea, of course, is that the unshared 2 percent is what makes all the difference.

It seems to me most critical to consider how biogenetic engineering will contribute to an increasingly domesticated world and to draw the lines for its implementation on those terms.

From what I have described above, wild and domesticated forms are both varieties of the "natural." Human ecologist Paul Shepard has discussed many reasons to value and preserve wild forms, which, of course, are different in profound ways from their domesticated counterparts. But from a strictly biological point of view, according to which even the human species, at least up till now, is "natural," I would counterpose wild and domesticated species on one side to genetically engineered forms, which I see as tending toward the status of artifacts. Advocates of genetic engineering claim that it is no different from what evolution has done, and that it is in fact a new form of evolution. But genetically engineered crops are not analogous to products of normal evolution. If epigenetic causation is the motor of evolution as I have proposed, and genes play a subordinate, consolidating role, then going at the properties of an organism by manipulating its genes is not even really "engineering." It is the hit-or-miss production of potentially useful monstrosities.

The current period is characterized by a growing drive to turn the living world into a collection of manufactured artifacts. Already the legal system says that if you make a genetic modification in an organism it's a human invention, it's not part of nature. This was the stated majority opinion of the U.S. Supreme Court in its 1980 Chakrabarty decision, which affirmed the right to patent organisms. I don't have anything against manufactured items, and will even acknowledge that genetically modified microorganisms may be useful. I use them in my own research. But I am dead set against patenting them. This takes the threat of blurring the distinction between organisms and artifacts that is implicit in genetic manipulation and turns it into a legal and cultural reality. The Chakrabarty patent was for an oil-eating bacterium. Since then it has served as a precedent for the issuing of patents on mice, pigs, and cows, some containing introduced human genes, as well as naturally occurring human bone marrow cells. There is no U.S. regulation that would forbid a patent on a genetically modified first-trimester human embryo—and such things would indeed be useful and commercially viable.

Do you see an irreversible threat to natural systems and evolution with transgenic introductions?

I do. There's a thicket of ideology that surrounds all of this that is important to understand. The biologist E. O. Wilson and his followers say that evolution is totally opportunistic, based on the harshest of organism-organism and organism-environment interactions, but, at the same time, the products of evolution are love-inspiring. They speak of "biophilia," our love for the living products of nature. Yet as the philosopher Hans Jonas notes, from a Darwinian viewpoint, evolution is nothing but the successive elaboration of "pathologies." In my view it is not enough to say that although life is the result of arbitrariness and opportunism, we should love it just because that's what we happened to get. Of course, many modern Darwin-influenced thinkers aren't as ardent as Wilson—they just think it's all meaningless. Another somewhat one-sided view of living organisms has arisen with applications of the mathematical field of complex systems theory. Although this approach seeks to identify living processes with dynamical phenomena neglected by genetic reductionism, it in turn ignores an organism's accumulated legacy of jury-rigged gene-based stabilizing mechanisms. If we look at a modern organism, we see that it is a composite system that bears the stamp of originating, self-organizing processes, but also exhibits the incredible integration and consolidation that results from vast periods of genetic evolution. As a result, the living systems that we are familiar with are very different from nonliving systems—even self-organizing dynamical systems.

How can we understand the question of what life is in a way that enables us to put biotechnology into perspective?

Biology, or at least biology as a traditional vocation—which is to understand what life is and how it works—is very different from biotechnology. Now the distinction has become blurred because of the commercialization of organisms, and because the ideology of the gene collapses everything into a single thing that can be sequenced, modified, bought, and sold. People too easily confuse the manipulations technologists can do for the types of things that evolution has done. Darwinists will say evolution isn't wise, it's just whatever works. I wouldn't want to anthropomorphize evolution and say it is wise, but neither is it arbitrary.

HOW WE CHANGED

Jerry Martien

first we made a genetically
improved cat

a cat with wings

before we knew it the cat had
eaten all the birds

so we had to make a genetically
faster flying bird

only it wouldn't
sing unless we gave it
money

so we made a genetic
dog
who'd sing whenever
someone told him to

we couldn't shut him up

the dog got on all the radio and
tv talk shows

logo
the famous genetically
engineered singing
canine

cloned a new
song for every day

and while the birds
bombed the cats
we all sang along with the dog.

DAVID LOY

REMAKING OURSELVES: A BUDDHIST PERSPECTIVE ON BIOTECHNOLOGY

As human beings, our greatness lies not so much in being able to remake the world as in being able to remake ourselves.

—Gandhi

But what does remaking ourselves mean? We can suppose that Gandhi was not thinking about genetic engineering when he emphasized its importance. Buddhism has also developed this notion, in ways that are more in line with what Gandhi had in mind. Whether or not Buddhism is a religion or a philosophy, it is a path that helps us remake ourselves in order to overcome our *dukkha,* our inability to enjoy our lives. (As the Vietnamese monk and Zen master Thich Nhat Hanh says, "Buddhism is a clever way to enjoy our life.") This requires conscious efforts to transform our greed, ill-will, and delusion into their more positive counterparts: generosity, loving kindness, and wisdom.

Now, however, we have access to a new way of remaking ourselves. How shall we evaluate it? For most of us, our excitement about this alternative way to reduce some types of *dukkha* (most obviously, by treating hereditary genetic diseases) is shadowed by worries about its dangers. In addition to the problem of accidents, which are inevitable, there is the question of who will own and control this technology, and for what ends. Given the increasing domination of market principles, will we end up with an elite of super-smart, super-beautiful, and super-healthy people who control the rest of us? Some scientists are already discussing the likelihood of this, and not all of them see it as something to be worried about. Given both its promise and its dangers, then, how do we evaluate such extraordinary new possibilities?

To a large extent, we do this in much the same way we have tended to evaluate most new technologies: by distinguishing between nature and artifact, between the "world of born" and the "world of made" (e. e. cummings), in order to take sides between them.

Our preoccupation with this dualism may be traced all the way back to the Greek origins of Western civilization, to its distinction between *phusis* and *nomos,* nature and convention (which, once recognized as a construct, allows the possibility of restructuring society and our natural environment). Much of the Western tradition can be understood in terms of increasing self-consciousness about their difference. Diderot, Rousseau, Herder, the Romantics, and such contrasted the organic and spontaneous with the artificiality and sterile rationality of convention; Kant, Hegel, Marx, Comte, and others were optimistic about our progressive capacity to understand and control the laws of our own development.

One could argue that the historical dialectic between them is the Western tradition, and as a result, our ways of thinking about such issues "naturally" (or unconsciously) tend to bifurcate into that dualism. In terms of bioengineering, this means that some people want to dismiss it out of hand as unnatural, while others cannot understand why we are so reluctant to take charge of our (otherwise random?) genetic destiny.

Those who yearn for nature usually evoke some period in the past, while those who are optimistic about our social and technological self-determination have high hopes for the future. As usual, nobody is satisfied with the present. But what motivates this dualism? That is a crucial issue if we want to escape the sort of deadlock that usually ensues.

I suspect this bifurcation has a lot to do with the conflict between two of our most basic human needs, security and freedom. This is more obvious individually: We feel a need to be free, but becoming free makes us more anxious and therefore more inclined to sacrifice that freedom for security, at which time we again feel a need to be free. We need to feel that we are unique, special in the universe, but then we want the security of being just like everyone else. "Between these two fear possibilities the individual is thrown back and forth all his life," says psychoanalyst Otto Rank.

In short, two of our primary needs—freedom and security—conflict. Is the same thing true collectively? With regard to the relationship between nature and culture, one basic issue is the meaning of our lives. To accept our culture as natural is to be grounded in the sense that our role in life is determined for us (often by religious belief), while the freedom to discover or construct our own meaning is to lose that security due to the lack of such a "natural" ground. Materially, however, something like the opposite is true: For most premodern societies, the physical conditions of their survival were often precarious, so we embrace technology to control and secure those conditions. The irony of the ecological crisis, from that perspective, is that our technological efforts to secure ourselves materially over the last five hundred years are what have caused the biospheric degradation that threatens our very survival.

Collectively as well as individually, then, we have tended to alternate between yearning for the security of a "natural" grounding and for the technological freedom to reconstruct the conditions of our existence. Each has its own attractions and problems. And today our ambivalence is complicated by the fact that, whether or not the technological genie should have been released from his bottle, he cannot be put back inside. Once human culture has been recognized as convention, you can't go home again, for an essential condition of those "closest to nature" is that they do not know they are close to nature. Nor would we want to return (even if we could) to a "natural" premodern society such as Tokugawa Japan, where hierarchical and exploitative political structures were presented as perfect because they conformed to "the order found in the manifold natural phenomena of heaven and earth." That is simply not an option for us.

So where does that leave us? If much of our thinking today about this new technology of bioengineering remains trapped within this dialectic, is there any other approach that might shed light on our ambivalence?

As a way to evaluate genetic engineering, I am suggesting, such a distinction between natural and unnatural is inadequate—and also, as it turns out, not very Buddhist. Shakyamuni's dharma did not distinguish them, and insofar as some such dualism may be read into his teachings, the Buddhist emphasis may be somewhat on the unnatural—for example, the celibacy of Buddhist monks and nuns. (Perhaps the only place where naturalness may be seen as privileged in Buddhism is occasionally in those schools influenced by Taoism or Shinto, such as Ch'an/Zen.)

In the same way, Buddhism provides no support for those who seek a return to some (preagricultural?) golden age in the distant past. The Buddha couldn't find the beginning of our *dukkha* and wasn't much interested in that anyway; his sole concern was to show us how *dukkha* could be ended. Nor did he express any interest in a technological solution to (part of) our *dukkha,* although such an approach wasn't much of an option for his culture anyway. You will not be surprised that I have not been able to find any references to bioengineering anywhere in the Buddhist scriptures, but it is clear that critiques which argue that genetic engineering is unnatural are somewhat unnatural to Buddhism. The Buddhist angle is different, although no less acute.

In short, our discomfort with bioengineering needs to be articulated in a different way than simply trying to privilege the nature side of the nature/technology dualism, and Buddhism offers us one such way. According to Buddhism, we are unhappy, and make each other unhappy, because of the three roots of evil: greed, ill-will, and delusion. As mentioned earlier, these need to be transformed into their positive counterparts: generosity, loving kindness, and wisdom. That is because actions motivated by negative intentions tend to bring about bad consequences, while actions motivated by good intentions tend to bring good results.

This implies that one way to evaluate this new technology is to look at the motivations behind our eagerness to exploit it. If this eagerness is motivated by generosity, loving kindness, and wisdom, we can expect it to bring about good results. We should be much more cautious if the motivations are wrong, because that will tend to increase our *dukkha,* not reduce it. Of course, it is not a simple issue of "either/or," for there is a wide spectrum between these two extremes; and both individually and collectively, our motivations are usually a composite of various factors.

Nonetheless, and despite all the public rhetoric about its potential to improve our lives (which cannot be denied), it is not difficult to find the three roots of evil motivating much of genetic engineering as it is being pursued today. The most evident is greed, in the form of personal desire for gain and corporate desire for greater profits. The recurring squabbles over patent rights for bioengineered life forms, and even for parts of the human genome itself, make that all too obvious. Since this problem has been so well exposed, there is no need to go into it any further, except to emphasize how this "root of evil" works to circumvent the kind of slow, long-term evaluation necessary for even the most innocuous and benign applications of genetic engineering. The rush to make big bucks from genetic engineering tends to make us all into laboratory rats.

The role of ill-will is less obvious, and one hopes less pervasive, but such motivations are present in the competitive pressures that often drive researchers eager for Nobel Prizes and corporations eager for lucrative patents.

The least obvious factor, however, may be the most problematic one: our collective ignorance or delusion. The issue here is what are the other possible motivations behind our preoccupation with genetic engineering. Do we really understand why this technology has suddenly become so important to us? Are we simply excited about new ways to help people (as well as make a buck), or is something else driving us? There seems to be a psychological law that when we are motivated by something unconscious, which we do not understand, our actions are much more likely to lead to unfortunate consequences. So it is important for us to be clear about all our motivations.

As one way to approach this, I would like to develop further the parallel between our individual and collective motivations to see what light this may shed on our general obsession with technological solutions to human *dukkha,* as well as bioengineering in particular.

From my Buddhist perspective, what is most striking about our collective problem today is how much it resembles the central problem for each of us as individuals: the sense of separation between an ego-self "inside" and an objective world "outside." For Buddhism, this subject-object dualism, which we tend to take for granted, is a delusion—in fact, the root delusion that makes us unhappy. It causes us to seek happiness by manipulating things in the objective world in order to get what we want from it, but that just reinforces our illusion of a self inside that is alienated from the world "out there."

According to Buddhism, this ego-self is an illusion because it corresponds to nothing substantial; it is *sunya,* "empty." Instead of being separate from the world, my sense of self is one manifestation of it. In contemporary terms, our sense of self is an impermanent, because it is interdependent, construct. Furthermore, I think we are all dimly aware of this, for this lack of self-grounding means that our sense of self is haunted by a profound insecurity that we can never quite manage to resolve. We usually experience this insecurity as the feeling that "something is wrong with me," a feeling that is understood in different ways according to our particular situation and character. Contemporary culture conditions many of us into thinking that what is wrong with us is that we do not have enough money; academics, like aspiring

Hollywood actors, are more likely to understand the problem as not being famous enough. But they encourage the same trap: I try to ground myself by changing the world "outside" myself. I try to subjectify myself by objectifying myself. Unfortunately, that is like trying to fill up a bottomless pit at the core of my being, because no amount of money or fame in the world can ever be enough if that is not what I really want.

According to Buddhism, such "personal reality projects"—these ways we try to make ourselves feel more real—cannot be successful, for a very different approach is needed to overcome our sense of lack. Instead of trying to ground ourselves somewhere on the "outside," we need to look "inside." Instead of running away from this sense of emptiness at our core, we need to become more aware of it and more comfortable with it, in which case it can transform from a sense of lack into the source of our creativity and spontaneity.

Now the big question: Is the same thing true collectively? Can this shed any light on our contemporary attitude toward technology in general? Individually, we usually try to solve our lack of self-grounding by grounding ourselves somewhere in the world—in the size of our bank account or in the number of people who know our name. Are we, then, attempting to solve our collective lack of self-grounding in a similar way, by trying to ground ourselves in the world—that is, by objectifying and transforming it? As Donald Verene says, "Technology is not applied science. It is the expression of a deep longing, an original longing that is present in modern science from its beginning. This is the desire of the self to seek its own truth through the mastery of the object." ("Technological Desire," in *Research in Philosophy and Technology* [London: JAI Press, 1984], vol. 7, p. 107)

What is that deep longing? Remember the problem of meaning that lies behind the nature/culture dualism. Despite their material insecurity, most premodern societies are quite secure in another way: The meaning of their lives is determined for them, for better and worse. In contrast, our freedom to construct our own meaning means we have lost such security due to the lack of any such "natural" ground for us. In compensation, has modern technological development become our collective security project?

Today we have become so familiar with rapid scientific and technological development that we have come to think of it as natural, which in this case means something that does not need to be explained. But in what sense is it natural to "progress" from the biplane to a space shuttle during one lifetime? (Bertrand Russell was already an adult when the Wright brothers first flew; he lived long enough to watch men land on the moon.) In response to our anxious alienation from a "natural" condition, we try to make ourselves feel more real by reorganizing the whole world until we can see our own image reflected in it everywhere, in the "resources" with which we try to secure the material conditions of our existence. This is why we can dispense with the consolations of traditional religion; now we have other ways to control our fate, or at least try to. But we must also understand how that impels us. Because the traditional grounding security of religious meaning has been taken away from us, we have not been able to avoid the task of trying to construct our own self-ground.

Insofar as there is something unconscious about this, however, there has been a distortion: No amount of material security can provide the kind of reality we crave, the sense of self-being we most need (which is, fundamentally, a spiritual need). Unfortunately for us, we cannot manipulate the natural world in a collective attempt to self-ground ourselves and also hope to find in that a ground greater than ourselves. The meaning of our lives cannot be discovered in this way. Our incredible technological power means we can do almost anything we want, yet the irony is that we no longer know what we want to do. Our reaction to this has been to grow and "develop" ever more quickly, but to what end? Our preoccupation with the means (the whole earth as "resource"), means we perpetually postpone thinking about ends: Where are we all going so fast? Or are we rather running away from something?

Another way to say this is that our technology has become our attempt to own the universe, an attempt that is always frustrated because, for reasons we do not quite understand, we never possess it fully enough to feel secure in our ownership. For many of us, Nature has taken over the role of a more transcendental God, because both fulfill our need to be grounded in something greater than ourselves; our technology cannot because it is motivated by the opposite response, attempting to banish that mystery by extending our control. Our success in "improving" nature means we can no longer rest peacefully in its bosom.

However, doesn't this approach smear all technological development with the same Buddhist brush? Such a perspective risks falling into the same pronature, antiprogress dualism that was questioned earlier. Furthermore, it is difficult to see how this offers any specific insights into what is distinctive about biotechnology and its ethical challenge today.

In response, it is necessary to emphasize that even if my supposition is true (and how does one evaluate something like that?), this does not imply anything like a wholesale rejection of modern technology. Remember, Buddhist emphasis is on one's intentions, for it is our problematic and confused motivations that tend to lead to the consequences we don't like. This allows us to evaluate specific situations by applying a Buddhist rule of thumb: Is our interest in developing this new technology due to our greed or ill-will, and, applying the third criterion of ignorance or delusion, can we become clear about why we are doing this? Among other things, this means: Do we clearly understand how this will reduce *dukkha,* and what its other effects will be?

Such questions encourage us, in effect, to transform our motivations in a way that will enable us to evaluate technologies in a more conscious and thoughtful fashion. One crucial issue in this process is, of course, who the "we" is. Transformative technologies have often been initiated without much thought to their long-range consequences (e.g., cars), and sometimes they have been imposed by elites with a firm belief in their superior understanding (e.g., nuclear power). The evaluation process I am suggesting would involve engaging in a much more thorough and wide-ranging discussion of what we collectively want from a technology. This will not stop us from making mistakes, but at least they will be our collective mistakes. Also, this will inevitably slow down the development of new technologies, something I see as being not a disadvantage but an advantage that will allow for a more painstaking scientific evaluation less subverted by desire for profit and competitive advantage.

What, if anything, does this specifically imply about biotechnology? In one way, nothing. Genetic engineering is not distinguished from other technologies in that they all need to be evaluated in the above fashion. But in another way, biotechnology is indeed special, because it presents us with the most extreme version (and perhaps the most extreme possible version) of the difficulty with technological objectification generally. The Buddhist claim that we are nondual with the world, not separate from it, means that when we objectify and commodify the world, we ourselves end up objectified and commodified, by it and in it. As the world has become reduced to a collection of resources to be managed, the physical and social structures we have created to do this have ended up doing the same to us, and we find ourselves increasingly subjected to them, as "human resources" to be organized and utilized in various ways.

Biotechnology signifies the completion of this commodification process. Life, or at least human life, has traditionally been seen as the last bastion of the sacred in our increasingly desacralized world, but now this last resistance to commodification is being overcome, and the category of "sacred" ceases to correspond to anything in our experience. There are many places and phenomena we cannot yet commodify, but in principle nothing now remains outside the scope of technological transformation into "resources" by a species that has yet to demonstrate that it is mature enough to exercise this power in a healthy way. Up to now, we have labored to ground ourselves by reconstructing our environment. Suddenly there is a new possibility: trying to ground ourselves by reconstructing our own genetic codes.

This raises in an especially acute way all the ethical and spiritual issues adumbrated above. If we allow the commodification of market principles the freedom that is now accepted in most other areas, this technology will become available mainly for those who can afford it, including wealthy parents concerned about the competitive advantages of their offspring, and wealthy individuals concerned to immortalize themselves with clones. In addition to this obvious challenge to more democratic aspirations, can such a technology be entrusted to a species unable to cope with its own death? To people who do not understand their own sense of lack—with historically disastrous consequences for them and those around them? Do we want the genetic code of life itself to become another hostage to our inability to accept the impermanent conditions of our existence in the world, to our inability to accept our own empty (nondual) nature?

In this fashion, biotechnology not only creates special dangers and possibilities, it prompts us to address all such technological issues in a more conscious and open way.

More concretely, what does this mean for how genetic engineering is being carried out today? From the Buddhist perspective I am offering, I see no way to escape the conclusion that genetic engineering, as it is presently being developed, should be suspended while we engage in a thorough democratic discussion of what we collectively want from it. The greed, competition, and delusion that motivate so much bioengineering research and application today are a recipe for disaster. Needless to say, we should have no delusions of our own that such a full public debate will happen unless we fight for it. The economic and political powers that be have too much of a personal stake in pushing it.

Nevertheless, it must be emphasized that none of the above implies that all genetic engineering is intrinsically bad. It does not deny the possibility that, sometime in the future, we may have economic and political conditions that will enable us to conduct it with more conscious and humble motivations: to reduce human (and perhaps other species') *dukkha.*

That is because the essential point of Buddhism is not to return to some pristine natural condition but to reduce our *dukkha.* Despite all the obvious dangers, there is the possibility that biotechnology may do that—for example (and most obviously), by treating inherited genetic diseases. Today, we cannot expect the CEO of Monsanto Corporation to be a bodhisattva, but we can hope that this may someday be a world where a much more cautious approach to genetic engineering will improve the human condition more than it will threaten it.

That, however, is unlikely to occur unless we also learn that it is more important to remake ourselves spiritually than to remake the world. And that brings us back to the issue of how we are to apply the Buddhist rule of thumb in evaluating technologies. Earlier, I emphasized the importance of becoming clear about why we want to develop a new technology. Now I ask: How can we become clear about this? In conclusion, I think we cannot evade yet another parallel between the personal and the social. As a society, we cannot expect to become sufficiently aware of our collective motivations unless we also make the effort to become more aware of our individual motivations. I suspect we will not be able to resolve our group sense of lack unless more of us individually address our personal sense of lack. If the root problem in both cases is due to our felt lack of reality, then we cannot expect wholly "secular" solutions to technological problems that, as I have argued, provide us with a spiritual challenge.

MARTI CROUCH

HOW TO BE PROPHETIC

I shut down my laboratory in 1990, convinced that our practices in agricultural genetic engineering would damage the common good—even though it took another five years for genetically engineered crops to become a commercial reality.

Now, after observing widespread consequences of genetically engineered crops, my early decision to cease and desist research is vindicated. We see an accelerated concentration of seeds, in a few multinational corporations, threatening world food security. Ecological problems are surfacing, such as toxic pollen and disturbed soil ecosystems, and we see genetic pollution, as pollen from engineered crops fertilizes nonengineered varieties and related wild species. The safety of genetically engineered foods has not been adequately demonstrated and labels are not required, making future epidemiological studies more difficult. Alternatives to further industrialization of agriculture are not being seriously pursued at universities and other research institutions as scientists put all their eggs in the biotech basket.

Looking back, what appears to have been a keen prophetic power on my part was actually a willingness to pay attention to the present. First, I saw that the methods and ways of thinking required for genetic engineering research on plants and life cycles were degrading to larger sets of relationships. From there, I simply extrapolated from the details of our day-to-day research to see the ultimate applications of that knowledge.

For example, the basic research on the molecular biology of seeds, which leads to the terminator technology, demonstrates the ways in which research conduct presages application. There is a formula followed by almost all scientists who wish to get grants for studying how seeds grow and develop. In nature, the life of an organism is best described as a continuous cycle from one generation to the next, but we break those cycles in our labs in order to study them. The seed cycle is represented as a series of steps, with one leading to the next: fertilization leads to the formation of a small spherical embryo, which then forms a heart-shaped structure, and so on, until the mature seed results. In turn, each step is represented as controlled by particular genes. Thus, the goal of seed research is to describe the kinds of genes needed for each step, then control and alter the seed cycle by manipulating those genes.

Most researchers break the cycle by either creating mutant seeds that stop developing at particular steps, or by using toxins that cause specific blocks in the process. Seeds are thus routinely soaked in carcinogens, mutagens, solvents, and inhibitors, or are subjected to X rays. We—my students and colleagues—made hundreds of mutant plants, with embryos and seeds that were grotesquely disfigured and incapable of continuing the cycle of regeneration characteristic of life. If we were clever, the genes from these mutants could be identified, cloned, studied, and altered. The genes could also be patented, bought, and sold.

Is it mere coincidence that the knowledge gained by breaking the seed cycle is now being used to construct seeds that cannot continue past a particular stage in their development, and used to keep patented varieties of plants from use by others? Several agricultural biotechnology companies have been awarded patents to genetically engineer crops that grow normally until they make seeds, but then produce toxins in those seeds to prevent them from being planted. This terminator technology is a kind of built-in patent protection system. Thus, basic research that treats organisms as machines with interchangeable parts and the life cycle as a series of steps controlled by genes that can be owned has led to an application with the same characteristics.

If the means predict the ends, then the smart thing to do is change the methods of gaining knowledge. I would like an agriculture that is whole and healthy and fully capable of regenerating from one cycle to the next, far into the future. That is why I decided to quit doing molecular research altogether and to learn about agriculture by participating in it rather than by studying it as an object. Any method that results in weakened, poisoned plants is unlikely to yield information of use in creating vibrant food.

It doesn't take a rocket scientist to figure that out.

ANDREW KIMBRELL

(Interview)

CRITIQUING THE BIOTECH WORLDVIEW

Casey Walker: In* The Human Body Shop, *you wrote, "Extending technology and commercialization to the living kingdom and body is among the most significant transitions in history." Will you begin by describing the importance of the larger picture, of systemic analysis, when we critique biotechnology?

Andrew Kimbrell: Yes. Systemic analysis of biotechnology, or any technology, is not easy and is rarely even attempted. In the past, our society has failed to ask the important questions about a technology prior to its widespread use. More typically, we allow a technology to become a routine part of our lives and then rely on regulatory agencies to work out the problems through so-called cost-benefit analysis, which usually takes place long after any rational limit to the technology is possible. This "too little too late" approach to technology has resulted in catastrophic impacts on the environment and on our society and culture. Unfortunately, this is the path corporations and government bureaucracies are currently pushing for the regulation of biotechnology.

What we fail to appreciate in our current approach to biotechnology is the extent to which technology is legislation. Deciding to use the combustion engine or nuclear power or biotechnology legislates our lives far more than most bills passed in Congress. The technologies we choose to implement determine much of who we are and what we do as a society. Technologies that destroy the natural world or encourage undemocratic and technocratic control of the basics of life should be vetoed like any other piece of pernicious legislation. Unfortunately, though we call ourselves a democracy, we do not have any mechanism by which we can vote on technology. The vast majority of Americans oppose cloning humans, the genetic engineering of foods, and much of the biotechnology revolution. But their voices are not heard.

Given that society is rarely given a role in technology decision making, how can we change policies on biotechnology or other important technology issues?

I'm a litigator and an activist. Though I write books and articles, I'm not an academic. I'm in the trenches. Most often this means filing suits or organizing campaigns to "stop the bleeding" caused by industrial technologies. We do want to stop as many forests from being clear-cut as possible, we do want to protect wetlands, and we do want to stop as many creatures from being engineered and released into the environment as we possibly can. It's very, very important to fight these battles, and I've devoted much of my adult life to it, as have many, many others. However, I also understand that this activist approach is profoundly inadequate in dealing with the problems we currently face.

Halting our indiscriminate destruction of the natural world, and the grotesque attempt of biotechnology to remake life, also requires that we change our collective consciousness and habits of perception in regard to nature. Until we change our reductionist views of nature we will continue to destroy it. A major priority for all of us, I think, is to become part of a new revolution of consciousness about life and the natural world.

Will you speak to how issues associated with biogenetic engineering should be articulated for the public?

I often say that biotechnology is taking the unthinkable, making it debatable, and then making it routine. Most people, when hearing that government researchers are putting human genes into pigs, or corporate scientists are cloning numbers of identical cows, are instinctively repulsed. There is great wisdom in this repulsion. It reflects an intuitive respect and empathy for the integrity of life and appropriate limits of human interference with the living kingdom. We need to affirm this popular repulsion against two powerful reductionist habits of thinking about life—modern modes of thinking that consciously or unconsciously are used by the biotech industry, academic science, and the media to break down public opposition to genetic engineering.

The first is the view that the life forms of the earth are little more than biological machines. This doctrine, called "mechanism," has been around since the days of Descartes. It is a very dangerous habit of perception. It glorifies efficiency over all other values. Efficiency is, of course, the optimal trait for machines. However, making efficiency the optimal trait for living beings is a pathology. Yet this is the goal of genetic engineers. When I have asked researchers why they are genetically engineering animals or plants, they almost always respond that the goal is to make these life forms more "efficient." Genetic engineering, they say, makes all of nature more efficient. What's the problem, they ask? Who's against efficiency?

I think, when taken to this extreme, we are all naturally repulsed by efficiency. I have never met a person who treats children, friends, pets, or any living being he or she cares about primarily on an efficiency basis. Only insane people would calculate how they can put minimum time and energy into their children to get optimal results in behavior or productivity. We don't treat the things we love efficiently. Yet the mechanistic mode of thinking has so dominated our minds that we rarely use the language of empathy or love when talking about nature, especially in the context of policymaking or in academia. But to defeat the eugenic drive of biotechnology toward making ever more efficient life forms, we are going to have to defeat the mechanistic worldview and substitute one of empathy and respect for life.

Beyond the mechanistic tradition, we also have the tradition of "self-interest" that Adam Smith and the early proponents of the free market decided was the principal incentive for human activity. The drive for personal gain has become, over the last two centuries, the basis for capitalism. This now global pursuit of ever more commodities has destroyed kinship, tribal relations, and spirituality. What's worse, our economic system has commodified everything. Everything is for sale as long as a profit can be made. The question that biotechnology and other advances in biology now present us with is the limits of commodification.

Are blood, organs, and fetal tissue commodities to be bought and sold? Can childbearing and children be sold through surrogate mother contracts? Can human genes and other body parts be patented by corporations? Clearly, creating this human body shop of biotechnology demeans life, our respect for one another, and results in unconscionable exploitation of those who, through poverty, are forced to sell parts of themselves. Stopping the habit of thinking of life as a commodity is why I've been involved in litigating against surrogate motherhood and the patenting of life.

In this sense, biotechnology, with all its perils, does offer activists the opportunity to stop the invasion of the market into the body commons and, by doing so, raises the vital question of establishing limits to the market system and the commodification of nature.

Will you describe the way you look at the systemic impact of biogenetic engineering?

The biotechnology revolution now is transforming agriculture, human health, reproduction, and even military weaponry. This widespread application of the techniques of genetic engineering is creating a growing number of environmental, economic, and ethical impacts. As for impacts on the environment and human health, one little discussed area of biotechnology is its use by the military. The United States has spent hundreds of millions of dollars in biowarfare research over the last two decades without any regard for the potential effects of a release of the deadly pathogens with which they are working in dozens of college and corporate labs around the country. We've won several lawsuits against the Department of Defense limiting their biowarfare research, but this aspect of genetic engineering and the potential for bioterrorism are the kinds of scenarios that keep you up at night.

More generally, the environmental risks of genetic engineering are not widely understood because of the failure of many in the environmental community to understand the nature of biological or genetic pollution. Most environmentalists have devoted themselves to the important issue of chemical pollution—a contamination model of pollution in which a variety of toxic elements are emitted into the air, water, and soil, causing contamination. Most environmental groups grew up in this country fighting for legislation based on the contamination model: the Clean Air Act, the Clean Water Act, FIFRA, TOSCA,* and so forth. Now, many of these same environmentalists are calling genetic engineering a "green technology" because it does not cause chemical contamination of the environment.

Environmentalists such as these completely miss the biological pollution paradigm, which is not a contamination model of pollution but a disease model. It's how we get sick. Right? Three people have a cold and I am exposed to them. I sit across the table from you and the bug goes to you, and so on. This is not a contamination model of pollution, it's a disease model. Some call it bioinvasion. The pollution of an ecosystem by a biological invader has, of course, already caused extraordinary devastation. We know how the chestnut blight and Dutch elm disease virtually wiped out those trees across the country, and how kudzu vine and zebra mussels have wreaked havoc. Now,

***FIFRA: Federal Insecticide, Fungicide and Rodenticide Act. TOSCA: Toxic Substances Control Act.**

through genetic engineering, we're creating hundreds of thousands of genetically engineered microbes, plants, and animals and releasing them into the environment. This represents a real biological pollution problem. In contrast to chemical pollution, which dilutes over time, the organisms involved in biological pollution reproduce, mutate, and disseminate. There's no control of them and you can't limit their spread.

Each instance of the release of a genetically engineered organism is a kind of ecological roulette. Scientists refer to such releases as low probability–high consequence risks. We could release a thousand types of genetically engineered bacteria into the environment and only a few will create a negative effect, but that effect could be catastrophic. One graphic potential example of the devastation involved here is a genetically engineered enzyme that breaks down vegetative matter into biomass. Many environmentalists see this as a wonderfully clean new technology that could reduce our need for fossil fuels. We're trying to substitute biomass for oil and gas and here's this little enzyme that breaks down vegetative matter with amazing efficiency. Well, what if that enzyme escapes? I admit, the oil crisis would be solved, but all our forests, crops, and vegetative life would be reduced to one large bog of biomass. It would mean the end of nature. That's biological pollution by a genetically engineered organism.

A subset of biological pollution is genetic pollution. This involves a genetically engineered organism passing on a deleterious genetic trait to a native species. A classic example: The Canadian government and several corporations have taken human growth genes and put them into salmon to create super-salmon. They've also inserted chicken genes to change salmon reproduction, so the salmon don't follow their life cycles up rivers. If one of these genetically engineered salmon is released—inadvertently or not, it makes no difference—we'll have a huge salmon swimming around with human genes and chicken genes in every cell of its body. This salmon will mate with wild salmon, and some of that human and chicken genetic material will pass over to the native salmon. The native genetic pool will be contaminated forever. And assuming that the greater size of the gene-altered salmon makes them easier prey or requires that they consume more than the ecosystem can sustain, the entire species could be compromised. Over the next decades, we could begin to see biological pollution matching chemical pollution as an environmental threat. It is essential that the environmental movement begin to understand biological and genetic pollution and make stopping it a top priority.

It's not difficult to see these blindspots as rooted in the difference, too, of a shallow environmentalism geared to the political wins of clean air, water, and food for humans, rather than a deep ecology with values for life itself.

Yes. And we can also see the biotechnology revolution as an extension of the pyrotechnology revolution. Since the Industrial Age, we've burned, forged, and melted the inanimate to create our metastasized nightmare of modern cities and factories. Now, from the industrial mind-set, we've simply added biology to physics. The only reason you and I find that so shocking is that we, just as in almost all traditional cultures, have deep regard for living things. The idea that we're going to use the same engineering principles of abstraction, quantification, manipulation, predictability for living things as we have done on inanimate nature is wholly shocking. The devastation that we have seen with the engineering of mountains, rivers, and soils is now becoming routine for all engineered life forms, including the human body.

Then, too, over the last two hundred years of patenting, machines have long since usurped our control over them. As social critic Lewis Mumford noted, the archway over the 1939 World's Fair carried the motto: "Science Explores, Technology Executes, Man Conforms." This was the religion of progress.

What an admission to slavery!

Yes, but initially a seductive one. When I was a kid, we thought we were going to become the Jetsons. Then suddenly, in the sixties and seventies, it became very clear that life as we knew it was not compatible with a technological system. We had so altered the biochemistry of earth with fossil fuels and global warming, ozone and topsoil depletion, that problems weren't localized but globalized.

Yet we have a whole technological system that has become the milieu in which most of us live and from which most of us relate to "life." It is not a natural milieu, nor the social milieu of agrarian societies, but a system thoroughly mediated by TV, markets, vast information and communication networks—all of which, as it turns out, is divorced from and destructive of life. In response, many of us worked hard within the "appropriate technology" movement. We read E. F. Schumacher, author of *Small Is Beautiful,* saw the wisdom of simplicity and small-scale economies, and believed we could devolve technology and devolve the system so that it would comport with ecological limits and not destroy, or perhaps only minimally destroy, the world.

Suddenly we are presented with another "solution," the solution of biotechnology. Which, instead of comporting technology to benefit life, asks, why not comport life so that it becomes technology? We've now come full circle with the complete invasion of technology—life actually becomes technology. We see this quite literally in the aim to download human brains into silicon chips. We see it in more subtle ways, too, in patented chickens and turkeys without the genes associated with brooding for greater efficiency at egg laying. Quite clearly, most agricultural research is working to make fruits and animals and vegetables comport better to factory farming. We're doing it! We're working hard to create weather resistance in a number of crops so they can survive global warming. In other words, large-scale efforts are now being made to comport all of life to the technological system. It's within that context that biotechnology can be best understood. We can all point to corporate greed, but in this larger, systemic context there's an impossible contradiction between a system of nature, or creation as it exists, and the technological system. One or the other has to give.

To a remarkable degree, we have already redefined life as machines and manufacturing under the patent system of the United States. We have done it legally. We also see a huge amount of computer literature referring to us humans as "biological machines." Our minds are software. As I mentioned earlier, this is the triumph of mechanism, the technologization, if you will, of life.

One of the greatest prides of modernity is that it is very pluralistic when it comes to religions. But the reason the larger system can be so generous is that this religious pluralism is irrelevant. Every religious tradition is completely irrelevant to the religion of modernity, which is the religion of progress, of the mastery of nature. So there is a default religion afoot. Make no mistake about it. We live it every day. It is the complete mastery of nature through technology to create convenience and wealth. I will go so far as to say that there's a new Trinity, a secular reductionist trinity, that mimics the sacred Christian Trinity. Science is God, all knowing, yet unknowable to most of us who didn't make it through reductionist college science courses; and it remains unknown, very mysterious. Science isn't too available or tempting, just like the God of the Old Testament, Yahweh. So, science incarnates through Technology, the Son. And the Son is with us. It's not abstract. Technology is the magic that works. How do we participate in the worship of Technology? We have the ersatz Holy Spirit, the Market. People don't get up in the morning for Science, and they don't get up just for the magic of Technology. They get up in the morning with the urge to make money so they can buy things. Filled with the spirit of the Market, we pursue upward mobility, more technology in our lives, and more consumer pleasures. Belief in the secular trinity is an essential dogma of the religion of Progress. Heretics to this religion are dismissed by scientists, businessmen, and economists alike. You tell us that science is going to let us know everything, technology is going to let us do everything, and the market is going to let us buy everything. That's progress.

Do you believe our best critique of biotech and modernity in general rests in our manner and motive of knowing?

Yes. You've brought up a good point, which is the violence we commit in relationship to the "other." I see this in the environmental movement and in myself sometimes. There's this sense that to know the other is in some sense to violate it, to push the other so you can know more. This often happens in relationships when we mistakenly think we're going to figure something out by pushing at it, pulling at it, teasing it, poking it to make it come out of itself. There's very little sense that the only way to know the other is to be in loving participation with it.

Do you view biogenetic engineering as an enterprise that not only denies but destroys limits of a natural and spiritual order?

It's the whole enterprise. The whole enterprise is to do two things: first, to make limits ontologically evil and drained of all meaning; second, to destroy any kind of sacramental imagination of life. Life can be seen as sacramental, even within, or perhaps especially within, its limits. The technological worldview looks on limits as ontologically evil. A recent AT&T ad promised a "world without limits where everything is possible." The goal of what I call the technological imagination is to destroy all limits and boundaries. Genetic engineering is now destroying even the boundaries and limits between species. The failure of the modern mind to appreciate limits is, of course, destroying the environment as well as indigenous societies and cultures around the globe. Traditional societies view limits very differently. They see them as meaningful, even sacramental. We need to regain this view of limits if we are to stop the juggernaut of biotechnology. In a sense, we need to create, or revive, a new sacred understanding of limits. This does not mean that we can give up the daily battles in courts and legislatures and in the grass roots. They need to be fought. But we will also have to attack the problem as its root, which is the technological imagination, and counter it with a revival of the sacramental view of life.

JACK TURNER

MODERN FORESTS

Critics of genetic engineering usually focus on the modification of humans, domesticated food crops, and farm animals. This should surprise no one given our anthropocentric culture. To the degree that anyone worries about wild species, they worry about ecological consequences of genetically modified crops, and until recently this has been only a minor concern. Then Cornell entomologist John E. Losey published a letter in the May 20, 1999, issue of *Nature* on monarch butterflies. Losey reported the death of monarch larvae that ingested milkweed leaves dusted with pollen from corn that had been genetically modified with the addition of *Bacillus thuringiensis* (Bt), a toxin that kills insects.

This raised obvious questions. How many other insects might be killed by genetically modified corn pollen? What about the pollen from other genetically modified plants, especially the genetically modified trees that are now the subject of research and development at prominent western universities?

Crops tend to be around other crops, but forests are where the wild things are—or supposed to be, or were. Have you ever seen the conifer pollen blow in western forests? It drifts in sheets of yellowish-green clouds, hundreds, sometimes thousands of feet high. It coats the lakes, deer, your truck, your barbecue grill. It floats down rivers. You inhale it, lots of beings inhale it.

Lots of critters live in the air, not just birds. According to an article by naturalist writer David Lukas in the spring 1999 issue of *Orion,* the "atmosphere over a single square mile of the earth's surface contains twenty-five million airborne insects"—everything from aerial plankton to flying spiders. What will be the consequences of Bt-laced pollen on all these insects? With transgenic forests, it is hard to see how wild beings can avoid eating Bt-bearing beings and transgenic beings. What will happen to overall trophic patterns of an ecosystem? No one knows.

Indeed, how many transgenic viruses, bacteria, mollusks, insects, birds, fish, and mammals have we created so far? No one knows—and no one is keeping track. What is documented is usually buried in arcane technical journals, and there is no "search engine" to collate the results of what graduate students are creating in the recesses of universities or what scientists are manufacturing in the secrecy of their corporate laboratories. How are we to get a handle on what genetic engineering is going to do to wild nature? After all, it is happening right now, and it will shape the future of ecosystems in the American West.

Consider one small area of research: common trees in western forests. Projects conducted at the University of Washington's Poplar Molecular Genetics Cooperative, Oregon State University's Tree Genetic Engineering Research Cooperative, the University of California, Davis's Dendrome Project, and at the Institute of Forest Genetics in Placerville, California (a unit of the U.S. Department of Agriculture, National Forest Service) have or soon will have manufactured transgenic versions of cottonwood (fifty transgenic lines!), aspen, willow, loblolly pine, Douglas fir, and western white pine. The goal of this research is stated clearly by the Institute of Forest Genetics: "To develop molecular and evolutionary genetics knowledge critical to maintaining viable populations of healthy, improved, genetically diverse forest species in sustainable western ecosystems. . . ."

This is to the point, but it veils what is really at stake: more efficient forests. One of those new transgenic cottonwoods grows ten feet a year.

This in turn veils the ways in which our ideas of "health," "disease," "improvement," and "diversity" are being modified by the concept of efficiency for the sake of greater profits. Transgenic forests are not about health, they are about money.

To produce "healthy, improved" species for our western ecosystems, research has focused on the same two areas that produced more efficient crops: herbicide resistance and resistance to insects and disease.

The herbicide is glyphosate, usually Monsanto's Roundup. Farmers spray Roundup on crops because it kills everything but the crop. Critics claim there are problems with Roundup; Monsanto denies it. According to "Roundup Ready or Not," an article published in *Seedling* (March 1997) and posted on the Genetically Manipulated Food News website, Roundup is the third most commonly reported cause of illness among agricultural workers in California, the top cause of complaint among landscape workers, and the most frequent cause of pesticide-related casualties in Britain. Roundup also blocks nitrogen fixation in plants, harms fungi, reduces winter hardiness in trees, and retards the development of earthworms. Do we want this stuff sprayed on western ecosystems?

Why spray a forest with Roundup and kill everything but the trees? What eats ground worms? How will this alter the phenology of the American robin? Since none of those trees will die from beetles and rust, there will be less material for wildfires. How will that alter wildfire regimes, landscape architecture? No one knows.

Now, as to the resistance to diseases and insects, they vary: cottonwood leaf beetle, tent caterpillar, and the fungal "pathogen" responsible for white pine blister rust. Resistance to disease is achieved in two ways: first, by a genetic transfer from another species—for example, by transferring a resistant gene from a sugar pine into a white pine; second, by transferring a biological pesticide to the tree. The choice in most cases—how did you guess?—is Bt, the same pesticide genetically engineered into Monsanto's corn, potatoes, cotton, and so on, to kill

Colorado potato beetles, bollworms, budworms, and corn borers, and the same pesticide that the corn pollen carried beyond the cornfields, dusting the milkweed leaves and killing the monarch larvae.

Aren't all those beetles, butterflies, caterpillars, fungi, rusts, borers, and worms somebody's lunch? Weren't genetic engineers required to study ecology at some point in their education?

Is anyone worried? Yes, the success of transgenic trees has created a new area of forestry: genetic risk assessment. It attacks the problem on two fronts. In one, Oregon State is working on a computer model they hope will predict the extent transgenes "will spread from transgenic plantations and what their impacts might be." Will spread? Might be? Does the computer model have a variable for robin survival and wildfire regimes? Why did they plant the trees before they had the computer model?

The other line of attack is sterile forests. If plain old transgenic forests are too dangerous, then we will have sterile transgenic forests—that is, forests where everything is either dead or incapable of producing life but very efficient at producing wood. This is a *reductio ad absurdum* of "forest."

Among the institutions supporting research on genetic engineering are Alberta Pacific Forest Industries, Boise Cascade Corporation, Champion International Corporation, Electric Power Research Institute, Fort James Corporation, Georgia-Pacific West, Inc., Inland Empire Paper Company, McMillian Bloedel Timberlands, Inc., Potlatch Corporation, Scott Paper Ltd., Shell, Westvaco Corporation, Weyerhaeuser Company, U.S. Department of Energy Biofuels Feedstock Development Program, British Columbia Ministry of Forests Research Branch, and, of course, the National Forest Service—the outfit that services our forests.

In short, the folks responsible for cutting down western forests now intend to replant them with either healthier, improved, more efficient transgenic goofies, or healthier, improved, more efficient, sterile, transgenic goofies.

Welcome to the modern forest.

As usual in forestry, the fusion of public and private interests is ubiquitous. In 1995, Potlatch and Boise Cascade were putting twenty thousand ha (about fifty thousand acres) of transgenic poplars "into production" under drip irrigation in eastern Oregon. At the Institute of Forest Genetics' Conifer Geonomics Research Program, half of the collaborators are from the Weyerhaeuser Forestry Research Center. When Oregon State planted several acres of transgenic trees on land near Boardman and Clatskanie, Oregon, in 1996, Fort James Corporation and Potlatch Corporation provided the land. It was, the university reported, "a shining example of university-private industry collaboration."

Unfortunately, there is nothing unique about genetic engineering for forests. In New Zealand they have successfully altered *Pinus radiata, Pinus taeda, Pinus elliotii,* and *Pseudotsuga menziesii* in "a high-tech quest to create forests of superior trees." In Brazil it is wild rice. In Australia, eucalyptus. In Chile, *Nothofagus alpina*—a hardwood endemic to subantarctic forests in Patagonia. India? Teak, moringa (a nut tree on the Malabar Coast), casuarina (a wood once preferred by cannibals on Fiji for making forks), and bamboo. Indeed, the Institute of Forest Genetics boasts that last year the institute hosted visitors from Finland, Germany, France, Spain, Hungary, Japan, Malaysia, China, South Africa, Australia, and Mexico. Soon we will have genetically engineered forests all over the world. How long will it take?

An article entitled "Poison Plants?" on the *Scientific American* website notes, "The percentage of genetically modified seed, some experts estimate, is now approaching 40 to 60 percent of all U.S. plantings." This took only five years. How long will it take for half the world's forests to be genetically modified? No one knows.

And should we be surprised by this endeavor? If we are ready to accept healthier, improved, efficient food and healthier, improved, efficient children, then what could possibly keep us from accepting healthier, improved, efficient trees, trout, elk—you name it.

If there is little doubt we will proceed to manufacture new forms of life in the face of ignorance, a veritable eruption of that most ancient malady, hubris, there is also little doubt that the gods will make us pay.

RICHARD HAYES

(Interview)

HUMAN GENETIC ENGINEERING

Casey Walker: Will you describe how you came to realize the significance of developments in human genetic manipulation and why you consider public involvement a matter of urgency?

Richard Hayes: As part of my dissertation studies at Berkeley, I wanted to learn about the new human genetic technologies and their social implications. I did course work in genetics and began attending conferences. I was stunned by what I discovered. We are very close to crossing technological thresholds that would change forever what it means to be a human being. The most consequential of these involve the modification of the genes that get passed to our children. In addition, there's human cloning, artificial human chromosomes, bovine/human embryos, "reconstructed" embryos using genes from three adults, and more. It sounds like science fiction, but it isn't.

These technologies are being developed right now in university and corporate labs, and neither policymakers nor the general public have any idea what's going on. These technologies are being promoted by an influential network of scientists and others who truly believe that they are about to usher in a new, techno-eugenic epoch for human life on earth. They look forward to a world in which parents design their children quite literally by selecting genes from a catalog. This would change everything we understand about what it means to be a parent, a child, a family, or a member of the human community. We would come to see people as artifacts, collections of parts assembled to achieve a particular result determined by someone else. Once we start genetically engineering our children, how would anything less than the "best" be considered acceptable? Once we start, where do we stop?

Until recently, these sorts of questions could be dismissed as speculative and far-fetched, but no longer. In 1998 a major conference was held at UCLA to promote the idea of how wonderful it's going to be when we can manipulate our children's genes and finally "seize control of human evolution." One thousand people attended and press coverage was extensive. Just a few months later, one of the noted scientists at the conference submitted the first proposal to begin experiments involving the modification of heritable genes. Things are moving very fast.

Mind you, some of these technologies hold great promise to relieve suffering and prevent disease. But we can draw bright lines to separate benign applications from those that are likely to set the world on a slippery slope to a horrific future.

Will you describe current genetic engineering technologies and those lines you believe can be drawn?

Sure. First, what's a gene? A gene is a string of chemicals that codes for and enables production of a particular protein, and proteins are the building blocks of our entire bodies. Genetic engineering is the process of adding, deleting, or modifying specific genes in a living cell. If your lung cells, for example, are missing a gene that produces an essential protein, you can use genetic engineering to try to acquire that gene. To do this you attach copies of the needed gene to harmless viruses, and let the viruses penetrate the cell walls and nuclear membranes of your lung cells. The needed genes are released into cell nuclei, incorporated into chromosomes, which are just long strings of genes, and, hopefully, begin producing the needed protein. That's genetic engineering.

However, an important distinction must be made between "therapy," which refers to gene modifications intended to address a medical condition, and "enhancement," which refers to modifications intended to improve some aspect of normal appearance or performance. Treating or preventing sickle cell anemia or cystic fibrosis would be therapy. Attempting to modify stature, agility, cognition, personality, or life span of a healthy person would be "enhancement."

A second important distinction must be made between gene modifications that have an impact solely on a single person and those that have an impact on a person's children and subsequent descendants. This is the distinction between "somatic" and "germline" genetic manipulation. Somatic manipulation seeks to change the genetic makeup of particular body (somatic) cells that comprise our organs—lungs, brain, bone, and so forth. Changes in somatic cells are not passed on to one's children. Germline genetic manipulation changes the sex cells—that is, the sperm and egg, or "germ" cells—whose sole function is to pass a set of genes to the next generation.

The critical question, and perhaps the most critical ever posed in human history, is, Where do we draw the line? Somatic gene therapy for individuals in medical need is already being tested, and few find it ethically objectionable. Somatic gene enhancement of people without medical conditions raises more concerns. Some somatic enhancements may be no more controversial than rhinoplasty, while others may be profoundly dangerous or otherwise unacceptable. But the effects of somatic enhancements are limited to a single person, so the risk to future generations is nil.

By far the most important issues concern germline engineering. Advocates of germline engineering invariably appeal to our compassionate desire to prevent the suffering often associated with heritable disease, but they're not putting all their cards on the table. Couples who believe they are at risk of transmitting a serious disease can already employ the far simpler technique of preimplantation screening to ensure that their children are free of the condition. In this procedure, a number of fertilized eggs are created in vitro—that is, in a petrie dish—and are tested to see which ones are free of the disease-causing gene. Only these are implanted. Any child subsequently born will be free of the disease, as will all of that child's descendants. The current aggressive push for germline therapy makes no sense, unless the real intent is to pave the way for germline enhancement, designer babies, and the technological reconfiguration of human biology.

Along the same lines, will you address human cloning and other technologies?

Cloning is the asexual creation of a human being by taking the nucleus from a cell of an adult or child and transplanting it into a woman's egg from which the nucleus has been removed. The resulting embryo would produce a baby that would be the genetic duplicate of the nucleus donor, similar to a twin. If someone cloned themselves, it's not clear whether the resulting infant should be regarded as the "sibling" or the "child" of the nucleus donor. In fact, it's neither; it's a new category of human relational identity: a clone.

Over the past century few issues have garnered such immediate and resolute consensus as has the issue of human cloning. Over 90 percent of Americans oppose human cloning. The great majority of industrial democracies, with the United States being the glaring exception, have already made human cloning illegal. Human cloning is condemned by every major religious denomination in the world. The United Nations, the G-7, the World Health Organization, and other international bodies have all called for a ban on human cloning.

Despite this, some scientists declare that they're going to do it anyway. Others say that although they are against replicative cloning—the cloning of fully formed human beings—they support the cloning of human embryos, which can be manipulated at very early stages to produce tissues for treating degenerative diseases. However, success in cloning embryos would make replicative cloning almost trivially easy. Further, the techniques of embryo cloning are precisely those necessary to make germline manipulation commercially practicable. This hasn't been mentioned in any of the media coverage of cloning. It's very difficult to get a desired new gene into a fertilized egg on a single try. To use germline engineering as a routine procedure, you'd start by creating a large culture of embryonic cells derived from a fertilized egg, douse these with viruses carrying the desired new gene, and transplant cell nuclei that have been successfully modified into new, enucleated eggs. These clonal embryos are then implanted in a uterus. Without embryo cloning, no commercial designer babies.

Currently, at least half a dozen approaches to producing therapeutic replacement tissues, none of which require embryo cloning, are under investigation. There's no overriding reason to develop human embryo cloning techniques unless the intent is to produce fully formed human clones or to make germline genetic engineering commercially practicable.

What is the significance of artificial chromosomes?

Germline engineering, in which the only goal is to change a single gene, is technically feasible today. But to engineer a child for more refined enhancements, many genes would need changing and current techniques are too crude. One solution is to build an artificial chromosome that contains all the necessary genes, organized in just the right way. Artificial chromosomes have been successfully tested in mice and in cultured human cells. The cells divide and the chromosomes are replicated intact. Now, human beings have twenty-three pairs of chromosomes and an extra, artificial chromosome pair would create twenty-four. If you wanted to have the benefits of the artificial chromosomes passed to your children, you could only mate with someone who carried the same artificial, twenty-fourth chromosome pair. One of the key characteristics of a species is that members of the same species can only breed with each other. So you see where this is going. In effect, we're talking about the possibility of creating a new human species, perhaps within one or two decades. Few people outside the science and biotech community are aware of this.

If the current pace of research and development continues, there will be an explosion of genetic knowledge and capability over the next several years. We will be able to transform the biology of plants, animals, and people with the same detail and flexibility as today's digital technologies and microchip enable us to transform information. The challenge before us is to summon the wisdom, maturity, and discipline to use these powers in ways that contribute to a fulfilling, just, sustainable world, and to forgo those uses that are degrading, destabilizing, and, quite literally, dehumanizing. Advocates of a full-out techno-eugenic future believe we're not up to that challenge. When push comes to shove, they believe, people won't be able to resist using a new genetic application if it looks like it might allow their children some advantage over other people's children. And they believe that once we allow even a little bit of germline engineering, the rest of the techno-eugenic agenda follows inexorably. I disagree with the first belief; I think we can be wiser than that. But I agree that if the germline threshold is crossed, further control becomes far more difficult.

The infamous slippery slope. Will you elaborate?

Suppose it became permissible to use germline engineering to avoid passing on simple genetic diseases like cystic fibrosis, even though preimplantation screening could accomplish the same result. What would the argument be against using germline engineering to avoid passing on predispositions to more complex conditions like diabetes, asthma, hypertension, and Alzheimer's—assuming the procedures were judged to be safe and effective? It's not obvious. After that, some scientists might offer gene packages that would endow healthy children with increased resistance to infectious diseases. Is this therapy or enhancement? It's a gray area. Similarly, what if genes that would predispose a child toward being very short could be engineered to predispose the child toward average height? How would you argue that such a genetic intervention be prohibited, assuming it was safe? Once it's accepted that parents have a right to use germline intervention to change a predisposition to shortness into a predisposition to average height, could you argue that they didn't have a right to predispose their child toward above-average height? Or toward above-average performance levels for a variety of simple and measurable cognitive skills? And after that, what about novel abilities that humans have never possessed before? Even if you banned such practices, advocates of germline manipulation say they'll just set up clinics in the Cayman Islands.

Scenarios like this persuade some people that resistance to the techno-eugenic vision is futile, and that we should just accept that it's going to happen. But think of the full implications. If a couple believes that it's desirable and acceptable to engineer their kids to be taller, wouldn't they typically also find it desirable to have a kid that's, say, less disposed to being overweight? Or disposed to being smarter, however they define that? Or more cheerful and outgoing? Or likely to live longer? Once you say yes to one enhancement, what rationale do you have for ever saying no to any other? If you accept that it's okay to engineer your kid, then doesn't not engineering your kid become something of a dereliction of parental responsibility? Especially when everybody else who can afford it is doing so? There are over eighty thousand human genes. How many modified genes do you want to put into your child? Ten? Fifty? Five hundred? Five thousand? Where does it stop?

Imagine explaining to your fourteen-year-old that you engineered her with a set of fifty or five hundred or five thousand carefully chosen genes. Now, imagine your child trying to understand who or what she is, and what's expected of her. Imagine her trying to figure out what about her is really her. Imagine her thinking about the children she would like to have someday and of the different ways in which she might like to engineer them.

Let's take it one step farther. Suppose you've been genetically engineered by your parents to have what they consider enhanced reasoning ability and other cognitive skills. How could you evaluate whether or not what was done to you was a good thing? How could you think about what it would be like not to have genetically engineered thoughts?

I think the entire scenario of genetic "improvement" is quite literally insane. The fact that so many educated, accomplished people seem untroubled by it is truly frightening. It's the materialist-reductionist-determinist worldview run amok. It's what happens when people become disconnected from themselves, others, and nature. I've been at conferences where participants use phrases like "when we start engineering our children," as if it's a foregone conclusion, with no indication that they appreciate the enormity of what they're saying.

In my opinion, there are clear lines that we can and should draw: no human germline engineering and no human cloning, ever. This is a moderate position, because it doesn't necessarily rule out many forms of somatic engineering or genetic testing and screening. We're going to have our hands full just deciding which nongermline applications to allow. Yet whatever we decide, we're not putting the future of humanity at risk; we're not eroding the basis of human individuality, self-regard, and autonomy; and we're not undermining the integrity of civil society and a democratic political ethos. But germline engineering and cloning, I believe, would set us on a path that leads in those directions.

I know some people argue that we don't need to be overly concerned about germline manipulation because, they say, it relies upon the discredited model of genetic reductionism, and thus will quickly be found to be ineffective. It's true, obviously, that the great majority of human traits involve complex interactions of genes, epigenetic biochemistry, environment, society, and free will. My guess is that over the next decade we'll find the full spectrum of possible relations between traits and genes: Some traits will be strongly influenced by genes, some will have little relation to genes at all, some will be influenced by genes in some environments but not in others, and so on. But in the absence of a ban, researchers will have no problem finding couples willing to run high degrees of risk in order to have a "superior" child. Some procedures will work and others won't. On balance, the techno-eugenic agenda would move forward. If we don't want to go down that road, we need to take stronger steps than, in effect, trusting the market.

Will you describe the world imagined by those advocating a techno-eugenic future?

The key text is Lee Silver's book *Remaking Eden: How Cloning and Beyond Will Change the Human Family.* It's one of the most pernicious books I've ever read. Silver envisions a world in which the new genetic and reproductive technologies are freely and fully used by everyone who can afford them, in order to give their children a competitive edge over other people's children. He acknowledges that this will lead to deeper class inequities, and then to a system of genetic castes, and eventually to separate human species, which he calls the GenRich and the Naturals. To those who want laws passed to ban the technologies leading to such a world, Silver sort of smirks and says, just try to stop us. He says that today's affluent professionals will develop and use these technologies no matter what the majority of people may decide.

It's difficult to overstate how grotesque a vision of the human future this is. It casually dismisses commitments to equality and democracy and common decency that men and women have struggled for centuries to achieve. It denigrates values of community and compassion as anachronisms ill-suited for the new techno-eugenic era. It celebrates nothing less than the end of our common humanity. Silver and his colleagues are quite aware of all this, but they really don't seem to care; they just want to enable people like themselves—smart, accomplished, aggressive, cynical—to get on with the business of segregating their "high quality" genetic lines from those of the rest of humanity.

It's astonishing that so few leaders in the scientific and biotechnology community have publicly denounced Silver's vision. I've spoken with many, and asked them to tell me how they believe his scenario can be avoided once we begin germline manipulation of any sort. A third of them avoid the question by making a joke. Another third say, "I don't know." And the final third say, "It's going to happen whether you like it or not."

Some people think scenarios like Silver's are so outlandish that they don't need to be taken seriously. I wish I could agree. It's important to remember that in Germany in the 1920s many people dismissed the Nazis as buffoons. Thresholds can be crossed that change realities of power and consciousness—we should know this by now. I'm not saying that techno-eugenicists are Nazis; in most ways they're quite the opposite, they're radical libertarians. Yet both are obsessed by the idea of the planned creation of biologically superior human beings. This obsession leads in only one direction. What would happen if the elites began engineering their children into a separate human species? There'd be protest, to say the least. Eventually the emerging GenRich would become impatient and start looking for a Final Solution. This is where the techno-eugenic vision leads. It's obscene and needs to be challenged.

Will you speak to the repeated claim that the techno-eugenic future is "inevitable"?

I think it's pretty apparent that claims of inevitability are rhetorical moves to rally supporters and demoralize opponents. Nothing in human affairs is inevitable. Most Americans are surprised to find that in the great majority of industrial democracies—all of Europe, Canada, Australia, and Japan, for example—both germline genetic engineering and human cloning have already been banned. The United States is the rogue country on these issues. The claim that people are incapable of agreeing to forgo individual, competitive striving in order to realize a larger social good is simply wrong. Of course, the fact that citizenship values are increasingly and profoundly being eroded by consumer values—in the United States and worldwide—presents a challenge. We're in a classic danger/opportunity situation: If we can't invoke and mobilize a sense of shared human citizenship, it will be difficult to constrain dangerous genetic technologies; on the other hand, the stark danger of these technologies might be just what's needed for the importance of a shared human citizenship to be widely understood and affirmed.

Some say that an authoritarian police state would be needed to enforce a ban on techno-eugenics, because people will do it anyway on the black market. That's hardly reason to accept and encourage it. Rather, we need to say with conviction that germline manipulation and cloning are unacceptable acts of power and domination by some persons over others, and we need to make clear that these technologies are not about curing disease, they're about turning people into artifacts. Strong moral suasion and effective laws can minimize and even eliminate black market abuses.

Techno-eugenic advocates believe they will prevail if they can convince people that bans on germline manipulation and cloning constitute infringements upon reproductive rights. We need to be clear that there's an enormous difference between seeking to terminate an unwanted pregnancy and seeking to manipulate the genetic makeup of a child and all subsequent generations. The great majority of people I work with on these issues support both access to legal abortion and bans on human cloning and germline manipulation. There's no inconsistency in holding both positions.

Will you give a brief chronology of the scientific developments that have led us to where we are today?

James Watson and Francis Crick figured out the structure of DNA in 1953, and by the late 1960s the genetic code for all the proteins had been deciphered. The ability to put genes into bacteria was developed in 1973, and transgenic mice were created in 1978. By the 1980s, proposals for genetic engineering of humans were being put forth, amid great controversy. A large coalition of religious leaders declared that germline engineering represented "a fundamental threat to the preservation of the human species as we know it," and should be opposed "with the same courage and conviction as we now oppose the threat of nuclear extinction." Germline engineering supporters decided to lay low and work instead to ensure approval of somatic therapy. In 1985, the federal government gave somatic therapy the go-ahead, and banned germline engineering "at this time." The ensuing race among researchers to be the first to "do somatic" was won in 1991 by W. French Anderson, who inserted genes into a young girl to treat an enzyme deficiency disease.

By the mid-1990s, articles began appearing with titles such as "Germline Therapy: The Time Is Near." In March 1998, the UCLA conference on Engineering the Human Germline was organized by vocal techno-eugenic advocate Gregory Stock. The event signaled the kickoff of a national campaign to, in Stock's words, "make it [germline engineering] acceptable" to the American people. The *New York Times,* the *Washington Post,* and other papers gave the event front-page coverage. A repeated theme was that germline engineering was all but inevitable. Stock said, "The question is not whether, but when."

After the event, Stock released a set of policy recommendations that called on the United States to "resist any effort by UNESCO or other international bodies to block the exploration of human germline engineering," and for the federal government to rescind its 1985 germline engineering ban. Three months later, the federal committee that oversees human genetic research, the Recombinant DNA Advisory Committee (RAC), discussed Stock's petition and agreed to review its policy on germline engineering. Simultaneously, the RAC received a proposal from W. French Anderson, the somatic therapy pioneer and a lead figure at the UCLA symposium, to begin a form of somatic therapy with a high probability of "inadvertently" modifying the human germline. It was an open secret that this proposal was a ploy. Anderson himself was quoted in the press saying that his proposal was designed to "force the debate" about germline engineering. If the RAC approves Anderson's proposal, it will establish for the first time that some forms of germline modification are permissible. As of today, Anderson hopes to be ready for human trials by 2002.

Will you speak to the challenges these issues pose for the environmental movement?

It's difficult to see how a world that accepts germline manipulation and the cloning of human beings will long be able to maintain, much less deepen, any sense of respect, reverence, and humility regarding the rest of the natural world. The techno-triumphalist vision calls for the wholesale transformation of literally everything living—plants, animals, humans, and ecosystems. It's not just a matter of putting a single pesticide gene into a corn plant or manipulating a single enzyme gene in a human zygote. What's under way is a reconfiguration of the deep structures of life. The new genetic technologies demand that the environmental movement deepen its critique if it doesn't want to be rapidly co-opted by an ecoutilitarian, technological worldview.

Have you heard of the new, transgenic EnviroPig? It's been engineered by Canadian scientists to contain both mouse genes and bacterial genes and produces manure with 20 to 50 percent less phosphorus than nonengineered pigs. It was developed to allow pork producers to raise more pigs per hectare and still comply with Canadian water-quality regulations. Should environmentalists feel good or bad about EnviroPig? Should we oppose EnviroPig but accept EnviroHuman? Or is it the other way around? Do we accept neither? Or both?

Here's another: Michael Rose at the University of California, Irvine has patented human genes that some scientists suspect might be able to increase our life spans up to 150 years. Should environmentalists oppose this, support this, or is it not an environmental issue? Students at UC Berkeley protested research on genetically enhanced life spans claiming it could lead to massive overpopulation and resource degradation. But if EnviroPig can alleviate water degradation, maybe we can engineer EnviroCattle and EnviroTree to alleviate other types of resource degradation. And after that, why not EnviroPlanet: a clean, green, nontoxic, nonpolluting, completely genetically engineered global ecosystem lovingly managed by genetically transformed EnviroHumans. This is exactly where we're going. Presently, environmentalists don't have a compelling way to say that this vision should be rejected. We really need to get to work.

Many are aware that the San Francisco Bay Area is now called the Biotech Capital of the world. Will you comment?

Genetic engineering proper started in San Francisco in 1973, when Herb Boyer at UC San Francisco and Stanley Cohen at Stanford figured out how to combine the genes of two different species. Three years later, Boyer co-founded the first commercial genetic engineering firm, Genentech. Today, the Bay Area has the single greatest concentration of biotech firms in the country. Besides Genentech there's Chiron, Shaman, Anergen, Clontech, SciClone, and many more. UC Berkeley just concluded a $25 million deal that gives the drug firm Novartis an unprecedented role in deciding UC's research priorities. In San Francisco, Mission Bay is being developed as a 120-acre biotech theme park. Of course, much of the research going on here is beneficial and deserves support. The problem is that the biotech industry is incapable, on its own, of drawing lines between what's acceptable and what isn't, and its increasing clout is enabling it to fend off attempts at regulation.

A critical case is that of Geron corporation, based in Menlo Park. Geron is potentially the ground-zero site for human cloning and germline manipulation worldwide. Geron recently announced that it had acquired Roslin Bio-Med, the firm that held the patents to the technology that produced the cloned sheep in Scotland. Geron has announced its opposition to replicative human cloning, and they're probably sincere, because there's very little money in it. What they really want is the freedom to clone human embryos and use them to produce replacement tissues for a mass market. Geron claims that it wants to find a way to produce replacement tissues without having to use human embryos. That would be a good thing; I support that. But get this: In 1998, Geron established an in-house ethical advisory committee of local bioethicists sympathetic to human genetic manipulation and asked their advice concerning human embryo cloning. The committee concluded that embryo cloning would be acceptable so long as the embryos were "treated with respect," which Geron promptly pledged to do. So Geron appears to be hedging its bets.

Have you heard that California has established an Advisory Committee on Human Cloning? It's dominated by the biomedical and biotech community and, incredibly, seems disposed to recommend that human cloning be allowed in California as an acceptable form of reproduction. This could be explosive.

What developments with implications for human genetic engineering can we expect in mainstream media over the next year or so?

Significant developments are going to appear in the press on an almost weekly basis. In early 2000, the sequencing of the fruit fly genome will be announced by Celera Genomics. Texas A&M hopes to announce the cloning of a pet dog, Missy, at a cost of $2.3 million donated by a controversial Arizona multimillionaire. Dr. James Grifo of New York University hopes to announce the birth of the first baby with genes from three parents, created as part of an effort to increase fertility among older women. Richard Plomin in the UK is expected to announce the discovery of multiple genes associated with IQ scores. The big event will be the completion of the rough draft of the sequence of the human genome in 2000, with the final version due eighteen months later. All these developments will be interpreted by the press almost exclusively through the framework of mainstream genetic triumphalism. At this time there are few effective voices offering an alternative, critical interpretation. As a result, some scientists and the biotech industry are controlling the development of public perceptions and public policy.

What is to be done?

We can take a deep breath and remind ourselves of the beauty and mystery of human life, and of all creation besides. Then we have to get to work. Germline genetic engineering is the single most portentous technological threshold in history, and we'll need a new social movement of commensurate scope and scale to prevent ourselves from slipping, or being pushed, over it. We'll need to alert, educate, and engage the general public, policymakers, and the press about what's at stake, and we'll need advocacy and political organizing as well. Substantively, we'll need permanent global bans on germline engineering and replicative cloning, at least a moratorium on embryo cloning, and an effective system of oversight for somatic genetic applications. We need to start talking about these things with everyone we know.

Educate yourself on the issues and figure out how organizations and networks with which you're affiliated can bring their influence to bear. The great majority of people recoil at the idea of humanity divided into GenRich and Naturals. We need to make it clear that the genetic transformation of human beings is something we neither need nor want to do. If we can accomplish that, we'll have established a new foundation for using our tremendous scientific and technological gifts in the service of a truly inclusive future for life on earth.

The GenRich—who account for 10 percent of the American population—all carry synthetic genes. All aspects of the economy, the media, the entertainment industry, and the knowledge industry are controlled by members of the GenRich class. . . . Naturals work as low-paid service providers or as laborers. . . . The GenRich class and the Natural class will become entirely separate species with no ability to cross-breed. . . .

Anyone who accepts the right of affluent parents to provide their children with an expensive private school education cannot use "unfairness" as a reason for rejecting the use of reprogenetic technologies. . . . Indeed, in a society that values individual freedom above all else, it is hard to find any legitimate basis for restricting the use of reprogenetics. . . . I will argue [that] the use of reprogenetic technologies is inevitable. It will not be controlled by governments or societies or even the scientists who create it. There is no doubt about it . . . whether we like it or not, the global marketplace will reign supreme.

—Lee Silver

Lee Silver is a professor of molecular biology at Princeton University. The above passages are from his book Remaking Eden: How Cloning and Beyond Will Change the Human Family *(New York: Avon, 1998).*

Many people love their retrievers and their sunny dispositions around children and adults. Could people be chosen in the same way? Would it be so terrible to allow parents to at least aim for a certain type, in the same way that great breeders try to match a breed of dog to the needs of a family?

—Gregory Pence

Gregory Pence is professor of philosophy at the University of Alabama School of Medicine. The above quote comes from his book Who's Afraid of Human Cloning? *(Lanham, Md.; Boulder; New York; Oxford, England: Rowman & Littlefield, 1998).*

And the other thing, because no one really has the guts to say it, I mean, if we could make better human beings by knowing how to add genes, why shouldn't we do it? I mean, sure, we have great respect for the human species. . . . But evolution can be just damn cruel, and to say that we've got a perfect genome and there's some sanctity to it? I'd just like to know where that idea comes from. It's utter silliness.

—James Watson

James Watson shared the 1962 Nobel Prize in Medicine for discovering the structure of DNA; he later served as director of the National Center for Human Genome Research and established the Human Genome Project. The above quote is from the website: www.ess.ucla.edu:80/huge.report.

By applying biological techniques to embryos and then to the reproductive process itself, Metaman will take control of human evolution. . . . Populations that adopt such techniques will generally outdistance those that do not. . . . Like all major developments, they will cause great stresses within society. But asking whether such changes are "wise" or "desirable" misses the essential point that they are largely not a matter of choice; they are the unavoidable product of the technological advance intrinsic to Metaman.

Even if half the world's species were lost, enormous diversity would still remain. When those in the distant future look back on this period of history, they will likely see it not as the era when the natural environment was impoverished, but as the age when a plethora of new forms—some biological, some technological, some a combination of the two—burst onto the scene. . . .

—Gregory Stock

Gregory Stock is director of the Science, Technology, and Society Program at UCLA's Center for the Study of Evolution and the Origin of Life. The above quote is from his book Metaman: The Merging of Humans and Machines Into a Global Superorganism *(New York: Simon & Schuster, 1993).*

> The right to a "custom-made child" is merely the natural extension of our current discourse of reproductive rights. I see no virtue in the role of chance in conception, and great virtue in expanding choice. . . . If women are to be allowed the "reproductive right" or "choice" to choose the father of their child, with his attendant characteristics, then they should be allowed the right to choose the characteristics from a catalog. . . . It will be considered obsessive and dumb to give your kids only parental genes. . . .

—James Hughes

James Hughes teaches at the University of Connecticut and is a bioethics consultant. The above quote comes from his essay "Embracing Change with All Four Arms," taken from the website: www.changesurfer.com.

> Chimeras or parahumans might legitimately be fashioned to do dangerous or demeaning jobs. As it is now, low-grade work is shoved off on moronic and retarded individuals, the victims of uncontrolled reproduction. Should we not "program" such workers thoughtfully instead of accidentally, by means of hybridization? Hybrids could also be designed by sexual reproduction, as between apes and humans. If interspecific coitus is too distasteful, then laboratory fertilization and implant could do it. If women are unwilling to gestate hybrids, animal females could.

—Joseph Fletcher

Joseph Fletcher was a professor of medical ethics at University of Virginia. The above quote comes from his book The Ethics of Genetic Control: Ending Reproductive Roulette *(Buffalo: Prometheus, 1988).*

CATHERINE KELLER

PLAYING GOD

I learned both what is secret and what is manifest, for wisdom, the fashioner of all things, taught me. For in her there is a spirit that is intelligent, holy, unique, manifold, subtle, mobile, clear, unpolluted, distinct, invulnerable, loving the good, keen, irresistible, beneficent, humane, steadfast, sure, free from anxiety. . . . For wisdom is more mobile than any motion; because of her pureness she pervades and penetrates all things.

—Wisdom of Solomon 7:21–24, M2SV

The question of whether we have arrived at a crossroads of "a world of made and a world of born" is beautifully put. One recalls that the very word *nature,* kin to nativity and natal, means nothing but endless birth. However, since I have agreed to respond as a Protestant theologian, let me say that the terms of the question suggest a dualism foreign to all biblical traditions. Since nature is understood as creation, it is all, in some sense, "made." Our mythology rotates around a God who creates creatures, who participates in the creative process; the earth and the sea are invited to "bring forth" those who will live within them; and the human earthlings (adamah, from adam, earth) are called "in the image of God" because we co-create in our linguistically enhanced sense—that is, by "the word." One can, as I and many other theologians do, free that notion of creation from its classical connotation of an omnipotent, anthropomorphic creator making "His" anthropocentric universe "from nothing." But I think Christian and Jewish theologies will not be able to draw any clear division between the made and the born, whether divine or cultural construction is meant: Ecotheologies in these traditions will need to draw the line not between the natural and the constructed but between the responsibly and the irresponsibly constructed. Indeed, Christian theologies that emphasize the pristinely "natural" or "natural law" usually do so in order to vilify homosexuality and abortion.

For us, the progressive path toward an earth-grounded spirituality will lie in deconstructing the nature/culture binary (the deleterious effects of its Christian forms) and not in romantically "returning to nature." So, if you think Christendom is the source of the problem—where else did technoscience get its aggressive confidence in its own universal truth, its drive to "dominion over the earth"?—I might not disagree. But if that means you think the solution to Western colonization of nature is to eliminate biblical metaphors, then, of course, you need read no further. My strategy must be to work within the myth, which presumes some human creative privilege within a nature understood as already created and creative, and to find its openings, its leaks, its porosities, and contradictions—the places where, for instance, it hints at the universal incarnation of the spirit in the flesh, and thus of the solidarity of all flesh, "members one of another." My strategy is to use our indubitable privilege finally—for God's sake—for the sake of the whole of life. Moreover, there is perhaps some tactical advantage in some of us claiming the biblical high ground, with its ancient tradition of protest against the commodification of life, to resist much of what profit-driven biotechnology has made.

As a representative of Protestant theology, however, I can't enter the discussion without first addressing a bit of bad theology used routinely by progressives, environmentalists, and other well-intended critics of irresponsible technoscience. This is the theology posted in the title of Michael Pollan's otherwise superb, indeed indispensable, essay in the *New York Times Magazine* exposing Monsanto's bioengineering of agriculture: "Playing God in the Garden." It is the theology assumed by the Prince of Wales, a passionate advocate of organic techniques, when he vowed that he would never eat or serve the fruits of a technology that "takes mankind into realms that belong to God and to God alone." "Playing God" seems to be the most charged public metaphor available for marking a point of transgression, a boundary of no return, an arrogance that inspires a kind of negative awe, a horror at some new level of Promethean hubris, some startling front in the technological colonization of human and other natures. It suggests that techno-science is replaying the myth of Eden, breaking the rules, violating a virginal paradise of nature, or perhaps defiling some final frontier of natural purity, such as its genetic integrity, threatening to get us all in big trouble (again?). It is because I am deeply concerned with that nature that I want to resist the bad theological tactics whereby we seek to protect it. They can only backfire.

Like "O God!," the metaphor of "playing God" is used by both non-theists and theists for its expletive force as well as its invocation of an absolute limit. Mainstream religious groups have used the concept with deeper theological consistency, as in the 1980 letter by Jewish, Protestant, and Roman Catholic spokespersons warning President Jimmy Carter against the attempt to "correct" our mental and social structures genetically. As theologian Ted Peters says in his book *Playing God,* at stake is the prescription of a "new commandment, 'Thou shalt not play God.'" Of course, the phrase has also served conservative moral ends, as in Christian opposition to such low-tech medical interventions as abortion and euthanasia.

Temporarily, however, my argument is not with the politics but with the theology. It implies an absolute boundary between the human and the divine, a boundary that we can and do transgress. This is paradoxical: God is usually depicted as the omnipotent creator; yet we have been granted the capacity to violate the creation (enough rope to hang ourselves with). "The earth is the Lord's and the fullness thereof!" exclaims the psalmist. But I don't think he or she had in mind the assignation of divine property rights implied by our capitalist discourse about "realms that belong to God." The phrase carries with it the implication not only that we are trespassing on God's property but that we are misappropriating divine prerogatives: whether of life and death or of genetic design. This implies that in some sense such interventions indeed imitate the work of God the creator. In the tradition of Protestant dissent against prior forms of Christianity, I lodge a double theological protest: against the twin assumptions that God can be separated from the universe, and that God is a transcendent agent who intervenes at will. This is the omnipotent God of classical theism, and, to be sure, this is the God to which the majority of Christians have been habituated. But "He" represents only one possible interpretation of scripture, and it is an interpretation that many Christians do not believe best serves the interests of responsible stewardship of the creation or the sustainability of our life practices.

But even if one holds to such a deity, as Peters more or less does, the phrase "playing God" inscribes a nonbiblical taboo against human technology. As Peters says in *Playing God,* "Science in the service of beneficence ought not to be intimidated by a 'No Trespassing' sign that says, 'Thou shalt not play God.'" How to determine "beneficence" is another matter altogether.

If we are going to talk about God (which I think Christians should learn to do as sparingly as did Jesus, who preferred "realm of God" or "kingdom of heaven"), then this God of "thou shalt not play God" doesn't wash. Getting human *is* "playing God." But play knows that it plays: It respects its own limits not by setting rigid boundaries but by transgressing creatively, lightly, knowingly. Play only turns sinister when it ceases to take its own power playfully and begins to stake out brutally serious claims. The degree to which science hardened into reductionist determinisms may reflect the degree to which it had to defend itself against theocratic suppression in its formative phases. That reductionism is beginning to loosen up in the theoretical sciences, at least as concerns quantum mechanics, chaos, complexity, and string theory.

Thus there is nothing more promising for the generation of an earth-grounded spirituality than its developing interchange with scientific cosmology. Perhaps this very interchange gives us the toehold we need for more precisely honed ethical challenges than "don't play God." Perhaps, indeed, it opens up a space for theological reflection. For as I will suggest, the ethical challenge is precisely how to critique the commodification of life. For me that means critiquing even the model of creation as divine commodity (God owns it, therefore you can't). And such a critique must rest in some form of constructive cosmology, some account of the relationship of our subjectivity to our materiality, of meaning to earth, that does not reinscribe either the old dualisms (mind over matter) or the old reductionisms (nothing but matter).

The great mathematician-turned-cosmologist A. N. Whitehead wrote in the 1920s that "religion will not gain its old power until it can face change in the same spirit as does science." By this he meant the spirit of adventure within an open universe of interactive events. The experimental sciences have been a major locus in the West of a highly disciplined but profoundly playful creativity, charged with a curiosity and wonder that the religious traditions have often stifled in themselves. Proponents of process theology, such as C. Hartshorne, John B. Cobb, Jr., and Marjorie Suchoki, have carried the Whiteheadian flame for decades: God appears no longer as the eternally unchanging, omnipotent designer but rather as the cosmic eros luring forth intensities surprising also to God. This God does not create as a means of control; the shape of creation depends entirely upon creaturely decision. Religion does not have to withhold some portion of reality from science for God. However, it does lay a godly responsibility upon humans—not just to make but to make good. When God declares the creation "very good," it is not as a king applauding his own accomplishment; it is the wild spirit of the universe, who, according to the Hebrew, "vibrates" over the face of the deep, delighting in its materializations. Clearly that delight relies upon and models an intensive sense of the common with extensive ecological, social, and economic implications (see Herman G. Daly and John B. Cobb Jr., *For the Common Good* [Boston: Beacon Press, 1994]).

When competition ceases to serve larger cooperative ends, the system begins to divide against itself. Responsibility does not mean obeying fixed legalities but, as the theologian H. R. Niebuhr had it, responsiveness in the old sense of fittingness (the nose is responsive to the face): responsive selves in relation. Responsibility intensifies creativity—scientific, artistic, emotional, religious—by the democratizing justice that enables intensification by mutual interaction.

Concretely, then? I don't think the ethical task is a matter of drawing an arbitrary line in the sand between acceptable and unacceptable levels of cultural interaction with nature. I don't think we can say in advance that all production of transgenic organisms, all creation of genetic hybrids, is irresponsible. We can say for sure, however, that the decisions as to which technologies to develop, how and for whom to develop them, are not being made responsibly. Given corporate control of funding for research, technoscientific agendas are being set for purposes of profit and power for the few. We must therefore create democratic channels by which we can demand that boundary-crossing biotechnical decisions be available to wide public scrutiny and debate, as is more the case already in Europe (partly spurred on by the campaigns against Monsanto's "Frankenfarms"). We are systematically being denied basic democratic consumer rights, such as the labeling of bioengineered foods, in the name of corporate rights. Moreover, the issues of race, class, and sexuality that will be arising in new and alarming ways along with the capacity for intruding in the human genome demand new means of public accountability. "Genes R Us" is quickly turning from a passive determinism to an active cultural determination of human "nature." The potentiality for humane cures to myriad human ills is attractive, but only to the extent the corporations dictating the technology are held accountable—no doubt through a combination of consumer pressures and government regulation—to a full spate of ethical perspectives.

I agree with biologist and science theorist Donna Haraway, who writes, "Genes for profit are not equal to science itself, or to economic health. Genetic sciences and politics are at the heart of critical struggles for equality, democracy and sustainable life" *(Modest Witness @ 2nd Millennium Female Man Meets Onco Mouse)*. But she goes on to sound the countertone, which ironically dovetails with the biblical position I have outlined. She criticizes the "left," which she identifies for its tendency "to collapse molecular genetics, biotechnology, profit and exploitation into one undifferentiated mass." However, I do not share the outright apocalyptic enthusiasm of my colleague Peters for what he sees as the technoscientific capacity to help God bring about the New Jerusalem, where "neither sorrow, nor crying, neither shall there be any more pain" (Revelation 21:4). The denial of death, which will always involve crying and pain, whether in theological or technological guise, will not contribute to the just and sustainable allocation of resources throughout the planet. The luxury of benign interventions at high expense for a few must always be weighed against the common good. For the majority on this planet, just six or seven decades of a reasonably well-nourished life would be pretty much heaven—and not far from what Jesus hoped for in the "new creation."

But for its misogyny, the more appropriate apocalyptic image for the current amalgam of technoscientific, transnational corporate power is Revelation's allegory of Roman imperialism, the Whore of Babylon. Thus, in the face of public protest in England, employees of Monsanto there have been, with the secret truth of the tongue in cheek, referring to their employer as "Mon Satan." Yesterday I saw a delightful one-act play by Daniel Kinch, *A Good Day to Pie,* whose solo role was performed by Rebecca Pridmore, a student from my seminary. The play, which exposes the frightening transgenic and transnational schemes of Monsanto, was inspired by the sentencing to a year in prison of a food activist who, after being repeatedly ignored, threw a pie in the face of a Monsanto executive. Why wouldn't justice be for sale along with life?

Since the biblical traditions gave rise to the notion of justice as liberation from oppression, it is against the commodification of life that the prophetic protest directs itself. "Justice is turned back, and righteousness stands afar off; and he who departs from evil makes himself a prey." But Isaiah, whose words remain all too relevant, is not entirely pessimistic: "If you pour yourself out for the hungry and satisfy the desire of the afflicted, then shall your light rise in the darkness." Indeed, the next verse suggests that we may need less fancy medical technology if we pursue justice instead: "And the Lord will guide you continually . . . and make your bones strong; and you shall be like a watered garden, like a spring of water, whose waters fail not" (Isaiah 58:11; 59:14).

To the degree that Protestantism has any moral voice to raise in response to biotechnology, it must ground itself in scripture. I opened, therefore, by quoting relevant scripture, for it is the enunciation of a form of premodern knowledge of the universe that builds into itself the criterion of beneficence. I must admit that, though this text belongs to the canon for Catholics and Eastern Orthodox, for Protestants and Jews it counts as apocryphal, deuterocanonical. This female voice of the Creator as Wisdom (in Greek, Sophia; in Hebrew, Hochma) has appeared again in our contemporary communities of discourse, prophetic precisely from her off-center but indelibly scriptural location. She appears more subserviently, but still luminously, in the canonical book of Proverbs as the cosmological Wisdom and "beloved daughter" (or female architect, depending on the translation) by which Yahweh creates. This cosmic Sophia, fashioner of all things, she who "orders all things far and nigh," provides the closest analogue to later natural law, or God's immanence in the creation. And at the same time, as the closest biblical equivalent to rationality, she is that by which humans participate in that cosmic knowledge. This does not mean we can expect to be wise. We may rather love and get wisdom; indeed, "wisdom is easily found by those who seek her." She does not play hard to get; but she does play hard—she "rejoices daily in the inhabited world" (Proverbs 8:30–31).

One role of the biblical traditions in relation to technoscience will be to call science to its own wisdom. We are not in a position to give or pronounce wisdom; we must, however, work incessantly to lure the culture of technoscience toward a form of knowing that goes beyond the pretense of value-free, objectively disinterested detachment, toward an intelligence that is "clear" and "penetrating" precisely in its humanness, love of goodness, beneficence, and justice. This can and must be cultivated from within science itself. Thus I take heart when I read biologist Stuart Kaufmann, who wrote in his book *At Home in the Universe:* "Since the time of Bacon our Western tradition has regarded knowledge as power. But as the scale of our activities in space and time has increased, we are being driven to understand the limited scope of our understanding, and even our potential understanding. . . . It would be wise to be wise. We enter a new millennium. It is best to do so with gentle reverence for the ever-changing and unpredictable places in the sun that we craft ever anew for one another. We are all at home in the universe, poised to sanctify by our best, brief, only stay."

To *sanctify* means to "make holy," and in such making, we are as a species and as its individuals both being born and being made, ever again. If God is holy, then of course we are, humbly and minimally, playing God. This play is constructive work. It is that for which we must take responsibility and through which we must try to slow technology down to the speed of wisdom. Wisdom in Proverbs invites all who would come to her table. Or, as *A Good Day to Pie* concludes, "This is the picture that counts. It's people sharing. A bowl of unadulterated rice. A bowl of beans that haven't been engineered. Breaking bread the way people have been breaking bread since there were people." Here's an odd hope: Perhaps with the decline of the cultural hegemony of Christianity, we may also expect less of the technoscientific defiance bred from Promethean revolt against a hegemonic God. In the subtle mobility of Sophia, permeating rather than controlling everything, there is little to react against but so much to respond to. New and beneficent coalitions may be forming just beyond this millennium, just beyond our knowing.

RICHARD STROHMAN

(Interview)

UPCOMING REVOLUTION IN BIOLOGY

Casey Walker: Will you begin by describing the properties of a scientific Kuhnian revolution, and why philospher Thomas Kuhn's understanding of paradigmatic shifts is key to a deeper critique of current developments in biogenetic engineering?

Richard Strohman: A scientific revolution is one in which a prevailing, dominant paradigm—one that defines a scientific worldview, together with the methods of achieving research-based understanding and technological utility—is replaced by a contending paradigm. The revolution, according to Thomas Kuhn, has several properties. One is incommensurability, where the scientists on either side of the paradigmatic divide experience great difficulty in understanding the other's point of view or reasons for adopting it. Second is the accumulation of anomalies, wherein "normal science" of the current paradigm unintentionally generates a body of observations that not only fails to support the paradigm but also points to glaring weaknesses in its method and theoretical outlook. The paradigm, under the weight of accumulated anomalies, loses the confidence of scientists in a minority sector and is ripe for overthrow. Third, paradigm shifts encounter resistance to change from the old guard, which is based not only upon a scientific incommensurability but on traditional ways of teaching and training the new generations in the (old) ways of research. Finally, there is enormous inertia based on the inability of a challenging paradigm to be fully capable of assimilating the accumulated anomalies and to provide a thoroughgoing scientific analysis and methodology capable of spelling out future programs of research and technology. The old paradigm may be a "scandal" of mistaken assumptions and failed predictions, but if there is no fully competent paradigm ready to take its place, the scientific establishment must, by definition, remain loyal to it. As Kuhn compassionately noted, scientists cannot at the same time practice and renounce the paradigm under which they work. Thus, because of incommensurability, resistance to change, and inertia, the discovered accumulation of anomalies will be ignored until a new and competent paradigm is capable of replacing it.

In the present climate of biogenetic engineering, which is based on the dominant paradigm of molecular-genetic determinism, Kuhn's critical analysis allows us to recognize features of the paradigm's relative success and failure over time. These features might not otherwise be noticed or, if noticed, might be forgiven for a host of reasons having to do with the belief that science is always an accumulative process, and, presumably, given enough time, any anomalies would be assimilated by progress within a "more of the same" pattern of research development. But in today's setting, where vast sums of intellectual and monetary resources and power are devoted to the genetic determinist paradigm and its application in fundamental biology, biomedicine, agriculture, and biowarfare, we need to exert every opportunity to examine our paradigm lest it be found, too late, to harbor anomalies that may turn out to be irreversible in the long run. Granted, it is not always a good idea to prejudge the performance of an ongoing paradigm, since it may indeed be self-correcting. But our dominant paradigm in biology has accumulated many anomalies, errant predictions, proven false assumptions, and outright errors of application in basic research and in biotech application in medicine and agriculture. The question is, Can biogenetic engineering in its present mode be dangerous to public health?

Kuhn's normal science in today's biology discovers its own flawed assumptions, but this does not lead to further insight. It reveals a complexity more grand than that imagined, but a complexity revealed is not the same as a complexity understood. The fact that we have discovered more than we understand—including the overwhelming databases of the various genome projects—suggests that our present paradigm is missing something essential. We need to recognize this incompleteness before going further with genetic engineering that will also produce unforeseen events in the earth's populations of animals and plants.

In "The Coming Kuhnian Revolution in Biology," you wrote "{W}e have taken a successful and extremely useful theory and paradigm of the gene and have illegitimately extended it as a paradigm of life." How did this illegitimate extension occur? Why did genetic determinism win out over systems theory, holistic biology, or areas of research in the complexity and nonlinearity of life?

The revolution in biology is all about the failed theory of genetic determinism. At the level of coded information in DNA—of replication, inheritance, and decoding of DNA messages—the theory of the gene is elegant in a simplicity accurately captured by what has come to be known as the central dogma of molecular biology:

DNA—>>RNA—>>Proteins—//—>>Functions

At this level the gene theory is complete or nearly so. However, this theory of what genes do and why genes are important has been extended to a theory of life which states that genes determine more than local function defined by individual proteins (and even this is exaggerated). Extended, the theory of the gene is a theory in search of ways in which genes determine complex functions, like normal or super intelligence, disease states, psychological states, and so on. It is this extension—from understanding inherited DNA as determinative of local protein functions to DNA as determinative of complex functions, not only of cells but of organisms of trillions of cells—that is illegitimate. It is this extended theory and paradigm of the gene that is in trouble, and where the revolution in biology is brewing.

It is easy to see how this illegitimate function was assigned. The structure of DNA as given by Watson and Crick was such a powerful insight about biological information, and generated such a productive wave of discovery and understanding at the molecular level, that there was every reason to believe that this information somehow extended beyond proteins to programs of behavior. However, no genetic programs have ever been found, and here is our dilemma. We organisms are certainly "programmed" in some sense of that word, but if the program is not in our genes, then where is the program? We have no answer to this question, and according to Kuhn, without a paradigm capable of addressing the mystery of the missing program, we will have to hang on to our incomplete paradigm of genetic determinism. Current references to nonlinear dynamics or complexity theory or chaos are all interesting starts at contributions to a new theory, but they remain as starts. That is our situation as we enter the new century. Our view of life is incomplete by half, at best.

For now, we adopt the shorthand expression for these alternative approaches and group them under the heading of "dynamics," the science that studies time- and context-dependent change in simple and complex systems. In life, both genetics and dynamics are essential. They are also irreducible, meaning that one cannot be derived from the other in any formal way. In life, genetics and dynamics are irreducibly complementary. In the last one hundred years we have had science based mostly in genetics, but mostly without dynamics. A science including dynamical systems offers many possible approaches in which genetics may be complemented so as to provide understanding of complex organisms.

Neither genes nor environments "cause" complex traits. If a word is needed here then "cell" will name the cause. It is the cell, and the body of cells as a whole, that selects from the dynamical interactions inherent in its physical and chemical pathways and responds formatively and adaptively to the external environment. We have mistakenly replaced the concept and reality of the cell as a dynamical center of integrative activity with the concept of gene causality.

Will you describe the anomalies discovered for genetic determinism? What doesn't genetic sequencing or complexity alone account for?

At all levels of life's organization—the evolution of populations, individual development, physiology, cell and molecular biology—and in applications in biotechnology and biomedicine experimental and field studies, we are discovering new facts that cannot be explained easily and sometimes not at all by the current theory. In many ways, these discoveries are merely modern versions of old but forgotten aspects of elementary (nonlinear) dynamic genetic processes, such as epistasis (gene interactions) and pleiotropy (one gene or protein with multiple effects). At all levels, one detects lack of correlation of genetic complexity with morphological complexity (an organism's structure and behavior). So, for example, differences in genome size and complexity between species are often much smaller than the differences in structure and behavior of those species (humans and chimps, for example, have DNA that is roughly 98 percent identical). Errant predictions based on genetics and the notion of specific cause and effect are turning up everywhere.

Development is a process in which identical genomes produce, in the case of humans, over 250 different cell types! In molecular and cell biology, one can delete genes with no apparent noticeable effect even though the deleted genes were thought to be essential. Once again, genetics alone fails to predict the correct outcome in these experimental settings. There are many other examples of anomalies.

On the other hand, complex dynamic processes do not account for the discrete informational bits in DNA and the syntax rules by which that information is encoded and manipulated. Genetics and dynamics are complementary, both are needed.

Is it possible that the scientific proof for a new paradigm will not come from any kind of determinism but from a description of motive, strategy, interactive will? If so, would theories along the lines of complex adaptive systems, epigenesis and nonlinear dynamics, chaos, or self-organized criticality be grounded in scientific proof?

We are at the beginning of a Kuhnian revolution, and, as Kuhn said, a scientist cannot remain a scientist and at the same time be without a paradigm. The result would be complete confusion. All or most of the candidate theories that I have gathered under the heading of dynamics do contain references to motive, purpose *(telos),* and will as needing to be included as essential irreducible facts of life, and all, in one way or another, address this necessity without needing to throw out reductionism or genetics, but to complement these with dynamics. It is much too early in the game to even guess how this is going to work out.

For now, pragmatics dictate a loosening of the theory of genetics to include dynamics. Biology needs more work, not less. While calls to abandon biotechnical applications based on incomplete science are not only pertinent but essential, such calls should not be misunderstood as being antiscientific—it is quite the reverse.

Will you explain epigenesis as an alternative to genetic determinism and describe the levels of dynamic organization it contends with? Further, will you speculate on the full import of understanding boundary conditions as key to the laws of living matter?

Epigenesis is the historical alternative to biological (genetic) determinism. It emphasizes a science of processes over objects, of order over heterogeneity and has, from the beginning, talked about the necessity of developing a science of qualities, purpose, and intentionality. The telos emphasized here has nothing to do with metaphysics or vitalism but everything to do with finding scientific laws unique to these qualities of life. Epigenetic process has been ignored for hundreds of years but has now been thoroughly revived—first as a necessary assumption to explain anomalies, and now as a description of regulatory (dynamical) processes that serve to regulate patterns of gene expression in a context-dependent manner. These networks are open to the world and provide the sought-after link between the environment, experience, and management of genetic information.

I wrote about boundary conditions as key to laws of life and I also spoke about epigenesis as able to provide only partial answers. Epigenetic processes are a class of dynamical processes operating in living systems. For example, they operate at the level of the genome to regulate patterns of gene expression—how signals from the world are integrated to turn genes on and off in an adaptive manner. But this is only one level of epigenetic regulation. There is a kind of infinite regress here since we now have to ask, What controls the control of gene regulation? The answer we might have to settle for is the cell. Then what controls the cell? And so on. Here we have to talk about boundary conditions. Michael Polanyi, the late, great professor of physical chemistry and of social studies at the University of Manchester (those were the days!) said the following: "Live mechanisms and information in DNA are boundary conditions with a sequence of boundaries above them" (see *Science* 160 [1968]:1308–12). He means that, in life, there is a continuous complementary relationship between genetics and dynamics from which comes the adaptive qualities of life. If we start with DNA, then the first, innermost, of a sequence of boundaries moving out to the cell as a whole is the epigenetic control of gene expression. This boundary moves outward to populations of cells, to the whole organism and beyond, to communities, populations, and, of course, to the world, natural or otherwise. They are all connected. The idea of boundary conditions gives us a scientific way of understanding the connections. As Polanyi put it, "The outer boundary harnesses the laws of the next inner one." To which one might add, "Or is it the other way around?" By partial answers from epigenesis, I mean that at some point we have to stop calling these levels epigenetic and call them something else. Polanyi's discussion reminds me of Wendell Berry's essay on the "system of systems" in *Standing by Words.*

What perspective do you hope society will gain, finally, on the usefulness of biogenetic engineering?

Biogenetic engineering, when faced with the limits and dangers of an incomplete genetic science, will have to conclude that engineering the human genome is, for 98 percent of our problems, facing an overwhelming complexity. That is the message from recently discovered anomalies and from experiments pointing to epigenetic, context-dependent regulation of the genome.

What then? Is that the end of genetics or of biological engineering? Certainly not. Biological—not biogenetic—engineering, through dynamics, will, or may, expand outward to include the organism and the natural world—the system of systems of the philosophers, poets, and scientists of the land, such as Wendell Berry, Wes Jackson, and many others. The human genome is ancient, conserved, difficult to improve upon most of the time and for most people. Biological engineering may use its databases not to engineer genomes but to understand the requirements of genetic agents and epigenetic processes in the world. Biotechnology and engineering would then be devoted to a restoration of the world to reflect the conserved genomes of organisms—human, nonhuman animals, and plants—that must occupy the land in common.

Biogenetic engineering assumes that organisms may be improved through genetic information almost exclusively. Biological engineering looks at the boundary conditions between genetic and other kinds of information and at the boundary conditions at all levels of living things, including the interface between the individual and the environment. As an example of the latter, elimination or reduction of human health risks (malnutrition, contaminated water and food, and tobacco smoking) has been the single most effective biological engineering in human history. We have already used it to gain thirty-five years of life expectancy just in the last hundred years. We have done this without engineering any genomes, but have instead provided those embodied genomes with a proper world, one that reflects their evolved limits and capacities. In fact, further improvements in life expectancy are expected not from eliminating diseases associated with old age, or from genetic breakthroughs, but from further improvements in eliminating environmental risks and in greater access to improvements for those who suffer premature morbidity and mortality. For example, people suffer from their lack of access to the food we can provide, not from our lack of ability to provide that food. However, as our world continues to suffer the disasters of our other technologies (land, water, air degradation), and as we continue in the futile hope of being immunized to the consequences of this suffering by genetic engineering—of ourselves or of the animals and plants—we will surely begin to lose the gains in health and life expectancy we have achieved this century.

In sum, several conclusions concerning the future of biotechnology seem inescapable. First, in the science of molecular genetics, the fundamental assumptions about specific genes and their specific "causal" effects on organisms are deeply flawed. As one recent paper in a major journal noted, there is now every reason to believe that it will not be possible to carry out genetic engineering (transfer of specific genes to a host cell) in the hope of achieving a specific effect. The normal complex interactions between genes and molecules in cells will be distorted by the presence of even a single transferred gene yielding unpredictable and therefore potentially dangerous results of unknown dimensions.

Therefore, biogenetic engineering of humans and of plants where unanticipated results could cause damage to individuals or to millions of acres of cropland will have to cease except under tightly controlled laboratory conditions and until the time when the complexities are understood and the dangers eliminated. Controls here would include concerns of ethical, legal, and social dimensions. These concerns must reflect the "ethics of the unknown" of the incompleteness of the science being applied, and not just the ethical concerns growing out of a "successful" technology.

There is, after all, an ethical component built into the structure of science itself, one that is often ignored by governmental and corporate structures as funders of research. This component includes the imperatives to seek evidence for disproving one's hypothesis (Karl Popper) and to consider all, not just selective, evidence (A. H. Whitehead). It includes also the historical record showing the capability of "normal science" to uncover the flaws and misconceptions of a prevailing paradigm (Kuhn). These imperatives and demonstrated long-run capacities are inconsistent with modern corporate technology, which is based on the need to produce marketable results in a cost- and time-effective manner and in a manner that deflects anomalies from consideration. Government support of "basic" research is all too often heavy-handed in insisting that all efforts be "sold" under the heading of being able to solve key problems or address other issues reflected, as Whitehead pointed out, "in the fluctuating extremes of fashionable opinion." Science is mostly a long-range affair while technology is not. The question remains: Who will pay for the long-range need to know?

Second, the flagship of biogenetic science is the Human Genome Project, and similar projects involving other animal and plant genomes are themselves squarely facing the anomalies discovered by their own scientists. Leaders of these projects are now increasingly aware of the flawed assumptions just discussed, and these projects are all now actively seeking to place their genetic findings within a wider physiological and ecological context. The search is on for the "meaning" of life that is now acknowledged not to be simply genetic in nature. The context has progressively shifted from epigenetic levels of genome control to a hierarchy of control extending to the cell as a whole and beyond the cell to cell populations, to organisms, and to the interactive and mutually dependent communities of life. The revolution against a preordained genetic determinism has discovered a complexity from which there is no turning back. However, without a fully developed paradigm of complexity as robust at the higher levels of biological organization as the Watson-Crick theory has been at the level of the gene, further progress will, predictably, be ruled by conservative forces. Therefore, in the interim, all caution must be exerted to guarantee that those forces also tightly regulate the old incomplete paradigm while we await its complementation with a science we have here called simply "dynamics."

This is the first time in the history of the life sciences that a single generation has been able to live through the rise and fall of a single dominant paradigm. It is a deeply disturbing experience, especially for those who have followed the radical change from a distance, and especially given the enormous investment our culture has made in ideas tied to a hopelessly ineffective, linear causality and determinism. Of course, from my perspective as a biologist, all this is wonderfully exciting: Science can and does work in a free society. As the great astronomer and cosmologist Sir Arthur Eddington has said of basic science, "We must follow science for its own sake whether it leads to the hill of vision or the tunnel of obscurity." Today, with the obvious link of science and society, there are many opportunities ahead—having to do with those as yet unexplored boundary conditions and mysterious spaces between hierarchical levels where "emergent" qualitative features arise from quantitative interactions below—to understand biology as the most complex of all sciences, and to understand the equally complex impacts of its applied technologies.

STEELHEADS, A PAGE OF EXTINCTION

Maya Rani Khosla

She enters the one pristine reach—
wetsuit, gloves, steel-toed boots slapping in,
angled over boulders, their medium
dark as another language, the words pushing in
on each other,
against her calves, that one torn patch in her suit
turned cold as gravel heap-hunched below color.

The river is a blue cave, she floats its palm, spine,
reading the slur, in it, the heaviness
dripping in from far fields like defeat,
the trickle of light along dawn's unrippled back.

Her own breathing is magnified,
pulled toward the rapids, sucked in, legs first,
one rock-grip to balance the pull
where water opens its soft ribs
and there beneath the gushing,
barely visible in a fog of debris
a whole school of them beat one pulse,
receding toward cover, bowing downcast,
like sad emperors of this ruined empire
their spotted fin-silk and noses bent
over stony worms.

FREEMAN HOUSE

(Interview)

WILD SALMON, *TOTEM SALMON*

Casey Walker: Will you begin by locating the Mattole watershed on the north Pacific coast and describing the physical place in which the events and experiences of* Totem Salmon *occur?

Freeman House: Cape Mendocino is the shoulder of northern California that juts out into the Pacific farther westward than any other landmass, and the Mattole River enters the Pacific about ten miles south of the cape. The Mattole is a relatively small river system that runs coastwise from south to north, about sixty-five river miles, and drains a watershed of some three hundred square miles. The Mattole also has the distinction of being located at the soft boundary of two bioregions, creating a sort of super-ecotone. The weather patterns share the regularity of winter-wet, summer-dry of the Shasta bioregion, but many of the vegetative patches are Cascadian. Some of the southernmost occurrences of red cedar, western hemlock, and Sitka spruce communities can be found here.

The Mattole sits at the junction of three large earthquake faults: the San Andreas (running southward), the Cascadian Subduction Zone (running all the way north into Canada), and the Mendocino (shooting out to sea, the most tectonically active place in North America). The King Range, which forms the western edge of the watershed, is rising out of the Pacific faster than any other place on the coast of the continent. It's rising at the relatively speedy rate of about fourteen feet every thousand years and does that in jolts and bumps. There's a place along the coast very near my home that rose up out of the Pacific five feet in a forty-five-second period in 1992. A very visceral way to experience the growth of mountains! As a result, the King Range and the watershed of the Mattole are very new lands. They're composed of soft, unconsolidated sea bottoms that are, even without interference by humans, eroding back into the sea almost as quickly as they're rising up.

Another important distinction is that the Mattole is home to one of the southernmost native stocks of chinook (or king) salmon on the eastern coast of the Pacific. Directly south is a two-hundred-mile gap of ruined habitat before other stocks, or races, of chinook migrate up through the San Francisco Bay and into the Sacramento or San Joaquin drainages. Mattole chinooks are one of the very last handfuls of native chinook salmon that hasn't been adulterated by hatchery introductions from other rivers. This gift of place made us realize that our job was not merely to increase numbers of salmon but to figure out a way to maintain that genetic integrity. An entirely different undertaking.

However, I use the Mattole as the source of my stories in *Totem Salmon* because the watershed offers a biogeographical scale of relationships and processes, of other species and histories, that not only satisfies my own personal longing for a certain kind of community and life, but also presents a configuration of inhabitation that is possible anywhere. The potential for rediscovering our roles as communities of place is, I believe, what the planet requires of us human beings. The salmon opened for many of us a pathway to engagement with the natural world. The scale of the watershed turned out to be just right: too large for any one person to perceive and comprehend, but not so large that the small inhabitory community couldn't begin to build a collective experience of it.

Will you describe the problems you found with declining salmon populations, and the significance of that restoration to the wild vs. hatchery salmon debate? How do your concerns reeducate those who see salmon crises in terms of fishing or grocery store availability?

First, we understood back in 1980 that our experiments in population enhancement were only a holding action, hopefully holding off the extirpation of the natives while we figured out meaningful ways to engage the recovery of the habitat and, ultimately, the recovery of cultural and economic practices that would resonate with the needs of salmon.

Second, as we know, the availability of anything in grocery stores has less and less correlation with the survival of the processes of the wild. By "wild" I mean complex systems that require no regulation or interference from outside themselves, that are self-generating, or autopoietic, and are demonstrated in the whole continuum of evolution to be the very basis of our success as living creatures. In the case of the salmon in particular, their co-evolution has been incredibly complex, elegant, and self-sustaining. Each stock of salmon has adapted itself over time to specific conditions of the freshwater habitat of its birth and early life and depends on those conditions remaining intact and diverse for the survival of the genera and species as a whole. When that diversity has been diminished to the point where only a handful of native stocks are left, it's not difficult to imagine a single exotic pathogen wiping out a whole species. What happens, then, to the co-evolutionary processes so evident between salmon and humans?

A kind of population crash we've seen recently with salmon in Washington's Puget Sound?

Yes. The insanity of rearing Atlantic salmon in the Pacific Ocean has resulted in the escapement of some of those pen-reared fish into the wild population and the setting off of epidemics of disease—diseases that hadn't previously existed in the Pacific—causing ecological diebacks. Only time will tell how long lasting those effects will be. Here is a great instance of strong global regulation being appropriate. There is no reason why anyone should be allowed to introduce Atlantic salmon and exotic pathogens into the Pacific Ocean. As I point out in *Totem Salmon,* hatcheries have been homogenizing the genetic diversity of salmon's place-specific adaptations.

Salmonid fisheries biology had been driven by hatchery technology throughout the last hundred and more years. The enormous numbers of fish that were, until very recently, being quite freely transported from place to place have, in many cases, overwhelmed the ability of native fish to compete. Hatchery fish are of the same species as the wild populations. They can interbreed, and they do. Thus they diminish and dilute the genetic adaptation that has evolved over tens of thousands of years. The growth of fisheries science is just now beginning to respond more to the needs of wild populations than to the needs of the marketplace, the engine that has tended, historically, to drive hatchery technology.

Will you describe the history of fishery management and the ways in which it has been driven by domesticating economics—by agricultural models—and recently by the sense that a "fish is a fish" when it comes to mitigation for compromised waterways and habitats?

Originally, hatchery culture came out of the same impulse that led to hybrid seeds for agriculture and domestication of the land for production of preferred foodstuffs. Now, "salmon ranching," where salmon are raised in large seawater pens until they reach a marketable size, is a logical extension of that same impulse. Early proponents of hatchery culture weren't motivated by mitigation for past damages so much as by the vision that they might increase the productivity of waterways that hadn't seen fit to entertain a large enough array of species. Fisheries biologist Daniel Bottom quotes (in "To Till the Waters: A History of Ideas in Fisheries Conservation," from *Pacific Salmon and Their Ecosystems* [New York: Chapman & Hill, 1997]) one of the most ambitious nineteenth-century advocates of the hatchery business, Seth

Green, in a statement that spells out his ambitions quite clearly. "We've been tilling the land for 5,000 years," Green wrote, "and we've only begun to till the waters" (Seth Green and R. B. Roosevelt, *Fish Hatching and Fish Catching* [Rochester, N.Y.: Union and Advertiser Co.'s Book and Job Print, 1879]). Hatchery culture has its origins in nineteenth-century entrepreneurial utilitarianism.

Fish stocks were introduced from one place to another all over North America and as far away as New Zealand and Chile. As wild fisheries were diminished by loss of habitat and overfishing, state and regulatory agencies picked up the idea of fish culture as mitigation for other economic activities that were overwhelming the West: dam building, irrigation, poor timbering practices, and on and on. At this point, many of the hatcheries that exist do so for purposes of mitigation of disastrous land-use practices that preceded them.

What is the current state of fishery knowledge and approach to both wild salmon populations and wild systems as a whole?

There has been a big shift in fisheries biology that has been driven, as best as I can understand it, by the American Fishery Society, the professional organization for fisheries biologists. In 1991, AFS published a devastating inventory of the number of wild stocks that have already gone extinct and identified nearly two hundred others that were in immediate peril. That paper has really created a significant shift in the way regulatory agencies manage stocks of fish. I think we can see a lot more care taken by hatchery managers in regard to their impact on wild populations, and a greater concentration by regulatory agencies on the health of wild stocks.

The fly in the ointment at this point—and that is much too small a metaphor for what's going on—is the development of large-scale aquaculture, especially among salmon. The fact is, most of the fish we consume these days is coming from "salmon ranching," where salmon are reared to marketable size in saltwater pens through their entire life cycle. It's a movement that started in Norway and spread rapidly to the west coast of the Pacific. So far, to its credit, California has resisted legalizing this activity. That is in part, I think, because California fishermen are well-organized, and fishermen are people who understand the value of wild stocks and genetic diversity and the higher quality of wild fish. This is something not to be undervalued. Anyone, and I mean really anyone, can do a blindfolded taste test of a wild fish side by side with a pen-reared fish and detect a dramatic difference.

We are also learning that attendant ecological impacts, both locally and biospherically, are dreadful. Locally, the concentrations of biomass from concentrated rotting food and fish excretions rob the wild waters of oxygen and cause local extirpations of other aquatic life. Disease runs rampant in such stressed environments, and when the penned fish escape, as they inevitably do, they carry those diseases into the wild populations. I try not to buy pen-reared salmon.

There's an obvious parallel between salmon and seeds in terms of control and homogenization. Much of the so-called promise of biogenetic engineering has been that it will solve food shortages, that we will be able to artificially manipulate and create our own foods to meet demand. This promise effectively marginalizes real damage to wild systems as irrelevant and brings about a whole new order of consequence, doesn't it?

Right. I think the largest intellectual argument of our times is just that: Can we exhaust the planet's natural provision and be clever enough to re-create it for our own purposes? It's a discussion that needs to be illuminated. It is the basic question at the heart of the survival of the wild and ultimately of our own species. There certainly is a lot of attention being paid to it at this point, but generally it's in the form of a discussion among specialists that is not getting out to people who are going to be dependent in part on either wild provision or a totally engineered world for the source of their food. The idea of putting our lives in the hands of a few experts rather than relying on the grand, self-generating, and self-perpetuating processes of the wild should make anyone distinctly uncomfortable. We have only to remember the promises of atomic engineers a generation ago to recognize the perils of such a course.

Will you describe the actual state of the land, the water, and the salmon as you found them in the late 1970s?

I was one of those people who came to the Mattole from life experiences that had been largely urban, and the place looked wonderful. It was green everywhere. The river has not been dammed, and to the uninformed eye it looks quite healthy in comparison to almost any river you'd find anywhere else in California. It takes some digging to find the differences between the way it looks now and the way it was as recently as forty to fifty years ago. The great eye-opener for me was to get my hands on the first aerial photographs ever taken of the place, in 1942, and to see the landscape as an unbroken sea of dark, old-growth forest so dense you could not find the river. In sharp contrast, contemporary photos show the river fully exposed for most of its length, with the vegetative cover much patchier. It's obvious the forests are much younger compared to what was here in 1942. In many cases, the original configuration of plant communities is not recovering. There is less prairie, less forest, more brushland and barren land.

About 75 percent of the land base is in forest soils, and three-quarters of the drainage is dominated by Douglas fir. Up in the headwaters, which for a variety of reasons has more fog, you begin to see the grandeur of redwoods.

By the time we arrived, more than 90 percent of the old-growth forests were gone, but what wasn't quite so evident was that the salmon populations were on a steep path downward. As long as humans had lived here, salmon had been a core subsistence food, and people who lived here were still acting as if that were the case. But, in fact, as we discovered in the first year or two of our activities, there were no longer enough fish in the rivers to support a subsistence fishery. That was one of the first things we had to take on. We were a bunch of Johnny-come-latelies who were trying to tell people who had lived here for several generations that it was no longer appropriate to do what they were accustomed to doing. It wasn't cheerfully received.

What experiences and ideas had previously shaped your and others' vision for the salmon and the community?

I developed my fascination with salmon nearly thirty years ago, while working as a commercial salmon fisherman. By coincidence, I also entered into conversation with social activist Peter Berg, California ecologist Raymond Dasmann, poet and essayist Gary Snyder, and others about a body of ideas we were calling bioregionalism. A bone-simple idea, in retrospect, that changed the course of my life.

I first became aware of Peter Berg's concerns when he returned from the 1972 U.N. Conference on the Environment in Stockholm convinced that the strategies of the environmental movement were a dead end. Its focus on global, political, and economic fixes for ecological degradation was not going to take us where we needed to go. He saw we needed a whole shift in the way humans perceived themselves as part of the planetary biosphere; the need for cultural strategies as a context for political struggles. It may sound like a huge concept, but devolved quickly, for me, into the realization that human engagement with ecological and evolutionary processes is limited by physical perception—by what we can see with our eyes, hear with our ears, and sense with our skins. Thus, the scale at which humans viscerally experience themselves as parts of the living biosphere is the key to our reorientation. Scale is everything when putting bioregional ideas into vernacular practice.

For quite a few years, the discussion of these ideas was largely theoretical. I quit fishing commercially for salmon by 1975, began researching and writing, and ended up in San Francisco not too long after the Planet Drum Foundation was founded. I worked in a little office in the basement of Peter Berg and Judy Goldhaft's house, helping to put out their publication *Raise the Stakes* and their first book, *Reinhabiting a Separate Country.* But I became impatient with the theoretical aspects of things and really wanted to get out into a particular place to test out the ideas and to see where they might lead. David Simpson, an old friend who lived in the Mattole, was becoming aware of the rapid diminishment of the salmon population. Suddenly, a piece of land became available that my partner, Nina, and I could afford to buy, and I had an opportunity to do what I wanted: to engage salmon directly in a community of place.

With this ideal in mind, what was the reality of the place and community when you arrived?

Nina and I finally settled in about 1980, and during the previous decade the population of the valley had tripled with people who were seeking a different sort of life from what had existed here before. It was a great immigration probably best described as the back-to-the-land movement. Due to a number of reasons—the breakup of two or three large ranches being one key factor—the Mattole had become a target destination for large numbers of people. As you might imagine, immigration on such a scale created a lot of social tensions.

People understood that they were seeing and taking less fish each year, but nobody was jotting down numbers. Regulatory agencies weren't providing the numbers because of the remoteness of the place. There are long-term cycles of greater and lesser abundance of salmon populations that are entirely natural, too. So it took systematic observation by local residents to demonstrate that the decline was indeed happening. We developed a protocol of counting spawning fish, applied that to a formula that was meant to give us a sense of what percentage of the fish we were seeing, and maintained it for twenty years. Even within a two- or three-year period, we were able to develop a pretty good comparative understanding of what was happening year to year. We were also beginning our work right in the middle of the greatest El Niño event in recorded history, and that was affecting the populations as well. In fact, once we were able to demonstrate those observations, people withdrew from their annual subsistence activities sadly but willingly, because there was a widespread recognition that it was not only a very precious resource but a whole definition of existence that no one wanted to lose.

Will you describe how the community's consciousness went from a sense of scarce resources and self-reliance to a deeper imagination of the place itself?

"Imagination" might be a better word than "remembering." I use the word *remembering* in *Totem Salmon* more metaphorically than literally, with the intent to speak to a continuity of human inhabitation of North America that it is a modern habit to ignore. We have a cultural attitude that history started when it began to be written down, which means when Euro-Americans arrived. The rapid elimination of the aboriginal culture here was simply more rapid than in other places, but no different in any other sense. Euro-Americans have never really had the intellectual tools to understand aboriginal inhabitation of the land; it's not so much that we have forgotten, but that we have never really known. We have "forgotten" that human inhabitation stretches back anywhere from one thousand to thirty thousand years, depending on where you are and to whom you are talking. That very important concept is seldom recognized, and it takes a vigorous application of imagination to inform the changes I see as necessary in our community practices.

I have come to think that engaging people with other species is the most direct way we have available to us to bring about transformations in individuals and communities. It doesn't have to be salmon that creates that shift, as it was in our case. If you look only at writers, Robert Michael Pyle is able to find a doorway into the natural world through butterflies; Richard Nelson through deer; Gary Nabhan through tepary beans, an aboriginal subsistence food in the Sonoran desert; Terry Tempest Williams through the birds of the Great Salt Lake. I'm convinced that once you find yourself inside the living system that sustains you, you will realize that the work to be done is only effectively undertaken in the context of the whole community. As an individual, you can take such attentiveness only so far. Gradually, if you're working toward the health of another species, you are led to the perception that the whole package of relationships is what you're attempting to rediscover. Any reading of the anthropological literature will bring you to the conclusion that this is the way humans have lived for most of our million or so years on earth. It's an aberration in our species behavior that we have lost social structures revolving around relation to place in the last five hundred to five thousand years.

When we don't live in relation to the larger system, we don't experience our individual and community lives fully. Just this morning I realized that the way I check out my moment-to-moment relationship to larger processes is by comparing it to those moments when I'm overwhelmed by the beauty of the landscape, oceanic feelings far beyond articulation. Almost everybody has experienced those feelings, but we tend to dismiss them as irrelevant, aberrant, orgasmic, and temporary. I am convinced they are signals pointing toward the fullness of life that is available to us. Perhaps this is what is meant by the traditional Navaho admonition to "walk in beauty."

Will you describe the greatest social challenges you encountered to creating this kind of community?

In hindsight, the greatest challenge was, and still is, our own ignorance as newcomers, something we share with most Euro-Americans. The only way to get beyond such ignorance is to systematically engage in learning about our relationship to the natural systems in which we are immersed, to engage all our senses directly with those natural systems and then to augment that information with what scholarship has delivered to us. The ignorance of which I'm speaking was compounded by our own countercultural arrogance. We moved into a new place assuming that we knew the best way for people to act. In fact, we knew very little about the people whose lives we were changing and even less about the ecological system that we were moving into and wanted so desperately to fit into. Most of us were identified in one way or another with the environmental movement over the past thirty years and had strongly held opinions and a contentious approach to the world. We were sometimes guilty of confusing the inhabitory people who had lived here a few generations before us with the corporations that were currently wreaking most of the havoc on the land.

It was an interesting time in the early 1980s because the California Forest Practices Act was just beginning to have its effect. Before that body of regulation, the private lands were in horrible shape and the public and corporate lands didn't look as bad. It was easy to be confused. Now, twenty-five years later, the situation has reversed: The small holdings, the noncorporate timberlands, are looking pretty good, and the public and corporate lands have been devastated. There's something very amiss in that picture. All of us together, the people who are engaged in watershed restoration and community building, some of whom take a living directly from the land and some who don't, are engaged in an ongoing, reciprocal dialogue, trying to understand the two (or more) communities, our common ground, and occasionally moving forward together. This proceeds in fits and starts because the whole back-to-the-land diaspora, which the entire West is experiencing now in one form or another, created resentments and fears and knee-jerk judgments in which people have tended to get stuck. We spend a lot of time unsticking ourselves, wherever we are.

What were the prevailing attitudes when you arrived?

The Mattole drainage is about 85 percent private property. About half of the remaining 15 percent is in corporate hands (and is managed with increasing aggressiveness) and half BLM [Bureau of Land Management] public lands, whose management style is increasingly benign. About 25 percent of the land base is in natural prairie, and up until about 1950 rangelands provided the economic base for the valley. It was largely a ranching culture here. With the whole variety of circumstances that emerged after World War II—including the development of technologies that would make these steep slopes available to chain saws, and a building boom that had recently discovered Douglas fir as an extremely valuable building material—suddenly that 75 percent of land covered in forest became tremendously valuable. At the time, regulation was ineffective, if not absent. Then, as now, whatever land-use regulations were in place were difficult to enforce. The result was the timber boom that wrought such havoc on the landscape.

Coming into that situation as I and many of my cohorts did, at the end of the timber boom, it was easy to confuse the people with the situation. The people we weren't quite displacing but were certainly disrupting, the people who were here when we arrived, God knows how they experienced us! We must have seemed like barbarian hordes. It's been a process of mutual and reciprocal education that we've all had to take on willfully to begin to understand that our commonalities far outweigh our differences. We'd all like to maintain some sense of self-determination in regard to our ability to endure here with some measure of self-reliance. By this time, the immigrants have been here for a generation. They tend to be the people who are inheriting the social and political infrastructure, and we're in a period of transition in that way, which isn't so different from thousands of other places in the American West. This is a phenomenon that rarely hits the news. The few publications that are devoted to proactive, place-based community development, such as *Chronicles of Community* and *Orion Afield,* have only appeared in the last few years.

One of the greatest strengths of Totem Salmon ***is its articulation of a place, a community, and a struggle that testifies to ecological ideals largely unknown in the urban or suburban mainstream, or even in the environmental movement as we've come to know it.***

The stories told in *Totem Salmon* could be stories told of thousands of other people struggling toward place-based culture, not only in North America but all over the world. It's also directly parallel with the struggles of aboriginal cultures that still have some remnants of their tradition intact. The whole thrust is to rediscover our participatory role. It may be institutionally impossible for mainstream culture to do that. It may be that "mainstream culture" is another way of misidentifying ourselves, as is the word *public,* as is the word *consumer.* When you work to perceive yourself as part of ecological systems that surround and support you, then those code words for social identity begin to fade in significance.

I'm not much of an intellectual, and I hesitate to make generalizations about most things, but the reason I'm so drawn to thinkers like Paul Shepard is that no one else I've read has articulated the passions that have driven my life as well as he has. Like anyone else, I am inundated with invitations to accept an identity that doesn't ring true and doesn't explain the profound sense of alienation I experienced in childhood and as an adult in our culture. What people like Shepard and, at this point, many others are illuminating is the difference between individualism—which has driven so much of our social, political, and philosophical thinking—and individuality, or the process of individuation. Thomas Berry calls "belonging" the opportunity to take on our true potential as human beings, which is to use our skills of articulation and our tools of rationality to celebrate the beauty of the processes of creation.

Will you describe examples of grounded and ritualized behaviors practiced by aboriginal peoples that reinforced belonging and the behavior of belonging?

I look for these examples, so I find them. In the ritualized behavior around technologies such as traps and nets and, in the case of the Ojibway, the snowshoes that allowed hunters to hunt in the winter, there is a sense of understanding that those technologies changed the relationship of the tribe to the world that supported it. We have lost that understanding in modern times, and we tend to feel we have no power over these technologies once they are launched, that we have no sense of choice about using them or about how they are used. That is something we desperately need to recapture.

What fascinates me most about these behaviors is the question of how those recognitions evolved: How did they come to be? In order to pursue such speculation, I was forced to use the various tribes who lived in the Klamath area because they still have remnants of traditional culture intact, along with a strong cultural revival movement, and there's more anthropological information about them than most aboriginal cultures in northwest California. It seems reasonable to me, given the knowledge that those various peoples immigrated into that basin at widely separated periods of time, that those ritualized intertribal behaviors of self-regulation that Euro-Americans found when they arrived did not fall out of the sky. They evolved over time. How did that happen? That's a useful thing to think about.

For urbanized or legally oriented societies, it's a radical shift to generate behavior from local knowledge and experience, isn't it?

Yes. If we look at the way our social and legal constructs are set up, we find no place for a voice for inhabitory rights, nor any place in the larger culture that provides an appropriate level of respect for inhabitory experience. I am fascinated with the evolution of communities. Again, scale is everything. I am not fascinated with the evolution of global culture, and the only resistance to homogenized, globalized culture that I can see lies in the evolution of place-based communities. How to instrumentally effect that evolution is a matter that deserves more attention than it is getting.

The word *ritual* gets overused and misunderstood a lot because it is, especially in the condition called modernity, rarely part of our secular lives. But after a few attempts at consensual decision making among people who, though coming from different cultural and economic backgrounds, can find commonality in their mutual inhabitation of a particular place, I've begun to think of that process as the sort of participatory ritual that might work. Consensual decision making has had a lot of bad press because it's compared to and perceived as undermining legal processes. This critique may or may not be true from situation to situation. But if a commitment to regular communication is understood as the process and practice of community building rather than as the goal-oriented technique for short-term decision making, then we can understand its usefulness differently. The appropriate analog for consensual practices is not politics but natural succession.

Commitments to the health of places are not going to be legislated into existence. They are going to evolve out of those places over a period of time that is probably longer than any one of us would like to think about. Gary Snyder says maybe in a thousand years we can realize bioregional goals; I'm a bit more optimistic than that. But it will be the work of several generations, at least. Consensual discussion does become a way to build community standards of behavior over time, but not in a legalistic sense. Such discussion is built on real conditions and phenomena, real experiences and needs, and on a basis that is directly understandable in vernacular life. Community standards are most effective and adaptive over time when they rise out of homegrown community ethics. Progress in that direction can be seen in subtle shifts in local attitudes toward land use, local food security, transportation, and above all in the growth of a kinder and more inclusive etiquette—how we treat each other and other species.

SALMON CREEK STEELHEAD
FOR RICHARD HUGO

Elizabeth Herron

He's coming down when I hear the splash,

see him dance above the cobble, spinning
into a crescent, an iridescent moon
at the curve of the creek where the roots
of alder hold the bank above the pool.

He drops back to water, moments later
shows up downstream, holding
in the current, lets himself slide
backward, fast and away.

But he was apples in a fog. He was fire
in winter grey. He was a god's glancing blow,
leaving a delirium of swirling water
pebbled with rain, wind in the empty trees.

KRISTIN DAWKINS

(Interview)

WTO AND TRADE POLICIES FOR GMOS

Casey Walker: Will you begin by describing the international trade agreements set by the Uruguay Round of GATT and the World Trade Organization, and how those have affected industries tied to biogenetic engineering?

Kristin Dawkins: The Uruguay Round was an eight-year negotiation from 1986 to 1994 between more than one hundred governments of the world to set new rules for trade. The GATT, or General Agreement on Tariffs and Trade, embodied trade law from 1948 until 1994, when the WTO was set up to manage the new Uruguay Round rules.

One of the issues that was not a high priority at that time, but is very much so now, is the trade of genetically engineered or modified organisms. Most often we think of genetically modified organisms (GMOs) as food, but the term also includes seeds that go into the production of food crops, and other products with altered DNA. The industries tied to GMOs are affected by three significant agreements set in those negotiations, under the auspices of the World Trade Organization (WTO): the agreement on agriculture, the agreement on food safety (officially called the agreement on sanitary and phytosanitary standards, or SDS), and the agreement on intellectual property rights, known as the TRIPs agreement.

Each has a different impact on trade and GMOs today. The agriculture agreement encourages trade of foodstocks, and will be renegotiated in 2000. The U.S. government has made it clear that high on its agenda for these renegotiations is to win agreement with the rest of the world that GMO foods will not be considered different from non-GMO foods. Indeed, our secretary of agriculture, Dan Glickman, has called it "the Battle Royale" of the twenty-first century.

Has the WTO's ruling, in favor of the United States, on Europe's ban on the hormone-treated beef set precedent for arguing the lack of difference between genetically engineered and nonengineered organisms?

It's a significant case and pertains to the food safety (SPS) agreement, which pertains to the health of the animals and plants that we eat. In that agreement, countries must justify their standards if they restrict trade or if they have higher standards than those internationally established by the Codex Alimentarius, which is the U.N. agency designated by the Uruguay Round as the official, international, standard-setting body for food safety.

The European Union's ban on beef imports and, indeed, on their own beef production with growth-inducing hormones, is considered trade restrictive. So, the European Union must prove its standards are "scientifically justifiable." This is where the dispute between the United States and Europe emerges—a dispute the WTO has been managing. The United States has charged that Europe's ban on beef products with hormones is not scientifically justifiable.

What would constitute "scientifically justifiable"?

It is a very subjective judgment. The WTO's appellate body said that Europe's risk-assessment process for this beef wasn't adequate or sufficiently grounded in sound science, and Europe was ordered to either lift their ban or conduct a new risk assessment. The European Union is preparing a new, science-based risk assessment for the WTO's consideration, but refuses to lift the ban. The United States argues that regardless of another risk assessment, Europe should lift their ban. The European Union offered to import organic beef instead, but the United States said no. It's been a long, tit-for-tat over the WTO's original decision. Now the United States has announced a restriction of European imports so that European producers of products other than beef will lose sales—producers of Danish pork, French cheese, and other chiefly luxury goods—as a way to pressure Europe to lift the ban.

From your point of view, is a "scientific risk assessment" legitimately possible in the public interest, or is it tied primarily to the interests of agricultural markets?

I would have to say that it is extremely political. I don't know the names of the scientists who have done the risk assessments, so I can't comment on them, but I can comment on the fact that the financial stakes are high and that representation of the biotech industry on the U.S. side—in our policy decisions, in our Department of Agriculture, in the White House, and in the Al Gore campaign—is all-pervasive. There was just an effort on the part of citizens across the United States to influence the National Academy of Sciences' panel on ethics to look at the ethics of biotech. They didn't have a single representative from what we'd call the side that "worries" about biotech. We objected to the panel's composition with a signatured petition and got them to appoint one "critic" from the Environmental Defense Fund, Rebecca Goldberg. She has quite a job there, one against many dozen, and we'll probably, at best, achieve a one-person minority-opinion report.

How is it that the Food & Drug Administration has ruled that bioengineered pesticides in crops such as potatoes don't require labeling because they're not considered food additives?

For one thing, the Environmental Protection Agency (EPA) regulates pesticides while the FDA regulates food additives, so there's a lot of passing the buck. Also, there's a huge revolving door between our government officials and the industry. This has been going on for a long time, but is especially easy to see now, with biotech, in this particular sector. The biotech industry has been in and out of the White House, the FDA, and today there's a leading person in Gore's office from Genentech. I think it does get back to campaign finance reform and time-out required between private- and public-sector employment, if we are to change the big picture for the long range.

In a larger context, there's corporate pressure at the international level. For example, due to campaigning on the part of the antiglobalization movement and a lot of effective citizen action in dozens of countries, it has been impossible to pass the Multilateral Agreement on Investment (MAI). But the worst part of the MAI shows what the real stakes are, because it gives rights to corporations to sue governments when governments issue regulations that interfere with a private company's right to make money in the future. Unfortunately, this set of rights already exists to some extent in the North American Free Trade Agreement (NAFTA), and there are a couple of cases seeking to prove them. The Ethyl Company, based in Virginia, makes an additive for gasoline (methylcyclopentadienyl manganese tricarbonyl, or MMT) that has been banned in the United States because of its carcinogenic properties but is still being sold in Canada. Last year, the Canadian parliament banned the additive. Under NAFTA's investment chapter, the Virginia company sued the Canadian government for what they'd lose in sales in Canada—something like $250 million per year—saying that if the Canadians thought it was worth keeping the ban, then the Canadian treasury, the citizens and taxpayers, would have to pay Ethyl for its losses. That was one of the most onerous parts of the MAI—these investment rights. When people find out about them, it's hard to explain to a thinking person why they shouldn't object. In France, campaigns have been especially strong, and the French government has ended up having quite an influence on the entire European Union.

However, there's another case, and we're back to the revolving door with another bovine growth hormone—in this case used on dairy cows to increase their milk production. Michael Taylor has gone back and forth between jobs at a private law firm (that counts Monsanto among its clients) and various FDA regulatory offices. While at the FDA, he was involved in a decision that approved Monsanto's use of rBGH for higher milk production in cows. There's now some evidence that this growth hormone causes health problems for consumers and plenty of evidence that it's a health problem for cows. Cows get infections and are fed antibiotics, and the antibiotics go into the milk, cheeses, and so on, and into our children. The public reacted, with a lot of campaigning going on in particular states because a number of farmers in those states were already suffering from low milk prices due to overproduction. Why would they want to produce even more? In the face of these objections, the FDA ruled that it wasn't appropriate for states to ban this product because at the federal level they had decided it was safe. A couple of state- and even city-level campaigns in New York and Chicago were deemed illegal by the FDA and upheld by the courts. This is the status quo in the United States.

Meanwhile, the Canadian government was trying to decide if it should ban rBGH and came to a quick decision when a health study conducted by Monsanto was leaked to the press. It showed that rats were indeed affected by ingesting rBGH. This study was never released in the United States. The FDA's policy requires safety testing only when recommended by the company. Monsanto acted within the law. Meanwhile, for a good decade, American kids have been drinking milk containing all this stuff.

Will you speak to the argument that biogenetically engineered crops and foods will solve world hunger?

World hunger, according to most analysts, including Nobel Prize–winner Amartya Sen, is not for lack of food. We have definitely been able to produce sufficient quantities of food, for decades, to feed the whole world. Productivity increases in food have been greater than increases in our population. It's much more clearly a problem of distribution, and this is political.

Today, the U.S. government is leading the charge for biotechnology by saying there is a huge fear that the world won't be able to feed itself in the future, and we'll have to use biotech methods to increase yields to keep people from starving. It is important to note that the yields they speak of are primarily from already highly polluted grain production regions in the United States and Europe, though they also speak of biotech in Africa and Asia, where it will supposedly increase people's capacity to feed themselves. We often note that even if food supplies were production-related, there are many means other than biotech to increase production.

The debate over food production is analogous to the old debate over energy production, about whether to take the "hard path" or the "soft path"—whether to build highly centralized nuclear power plants or to invest in decentralized, small-tech methods as better and cheaper in the long run. Biotechnology is the "hard path" with a very high-tech route versus a lot of investment in small farms throughout the world to enable countries, regions, even valleys, to be more self-sufficient.

At the World Food Summit, which the United Nations sponsored in 1996, there were hundreds and hundreds of nongovernmental organizations, ranging from think tanks to farm organizations to technical assistance kinds of groups and so on. Almost all of them were unanimously in support of the soft path of small farms and appropriate technology. The United States spent all its time arguing that the hard path was the only way to feed the world. The American delegation is very hardheaded, and they have the power to say no because they are the hegemonic institution in the world today. Of course, it's very ideological.

Will you describe that ideology?

It believes the market is the salvation of humanity. If we subsume all other social, environmental, and cultural interests to the market, then, in the long run, we'll all be better off. On the other hand, there are those of us who say the market is a very fine and effective instrument, but, in fact, there is more than one market, and they include local valley markets in Peruvian highlands where peasants trade potatoes from one valley to another. The conviction that there is one holy market that will solve the world's problems is not based on anything but an ideology. If you look behind that ideology you can see the clear, economic interests of the Fortune 500, the biggest companies in the world. These companies see the planet as their market, and if they can convert all of the planet's citizens to being their consumers, of course their profits will rise. You can find some people in government who see through this, but most are acting in the interests of these corporations, and there are still others who simply believe in the theory.

Will you speak to the idea that many of the policies protecting and promoting this ideology are in violation of the Universal Declaration of Human Rights?

Yes. The Universal Declaration of Human Rights, developed by the United Nations in 1948, defines and defends human rights as including the right to food, shelter, education, and quite a number of things basic for life. Then, we also have defined "industrial rights," with the theory that if industrial rights are well-defended, such as with the MAI, then the welfare will trickle down to the rest of us. What we find in practice is that welfare doesn't necessarily trickle down. Welfare has everything to do with government policies and nothing to do with the market itself. In the meantime, human rights are being abused right and left by this market theory.

This brings us to the third WTO agreement, which is on intellectual property. It says that governments must allow for patenting of intellectual knowledge. Back in the beginning of intellectual property rights, in the Jeffersonian period, there was the belief that those of us who use our intellect to invent things, to write or make beautiful music, or create other works that enhance society, need to be given some kind of share in the commercial reward thereafter. Copyrights on books and music and so forth are all fine and good, but the intellectual property rights agreement of the WTO, which is the TRIPs agreement (Trade Related Intellectual Property Rights), actually requires governments to award patents on living things.

This is where, in many cases, the knowledge of farmers and indigenous people that has gone into improving plants for foods or medicines or whatever for hundreds of years is suddenly being privatized by whoever files for and gets the patent. That patent holder has the right to exclude everyone else from its use for about twenty years. Patents have now been applied to quite a variety of seeds and plants. The exclusivity of it is very much in violation of the human rights of the rest of the world. There is an international campaign to try to overcome this 1994 TRIPs agreement and to inject the idea that patents on life may be prohibited. We'd actually like to see such patents prohibited universally as an international human right, but the politics have not been encouraging for that at this stage.

The *sui generis* option is key. It means, in Latin, "of its own kind." TRIPs allows governments to legislate *sui generis* rules to protect plant breeders' interests. There is some indication that the United States and a few other patent-holding countries would like to eliminate this provision. We need to insist that countries retain the legal discretion over how they protect farmers, indigenous peoples, and others who have used their intellects to improve plants. It allows for alternatives to patents, and is the key to allowing governments to implement another international treaty: the Convention on Biological Diversity.

Does patenting create an incentive to genetically engineer in order to "own" any given seed, plant, animal, and so forth?

I think you could argue that it does, though certainly it's not the only driving force. There's another part of the TRIPs agreement, unlike the patenting or *sui generis* protection of plants, that says governments must create a patent system for genetically engineered organisms. There's no way around it, which creates, as you suggest, an incentive for companies to take what is the raw material from some other country and, by moving just one gene around through genetic engineering techniques, makes it automatically eligible for a patent. If they hadn't moved the gene around, then the originating country could implement its *sui generis* option to protect the material for their national interests. I do think this incentive exists.

Recently, I read an article saying that in West Africa they had found a fungus that has properties much like insulin and could replace the insulin shot that diabetics need. For diabetics that's an attractive alternative. Now, if I were that Congolese government, I would want to implement the *sui generis* option that meets the TRIPs requirement, saying that the fungus is found to be socially useful and we want to retain our rights to not allow an exclusive patent for one company to do research with it, but rather to encourage its broad public availability. I'd argue that with more companies doing research, we might be able to come up with something that might work sooner; that it would be cheaper as a generic drug for consumers, so that diabetics all over the world would have better access to this new drug; that the commercial royalties should be shared with the government, which would in turn share it with the people who live in that region; and finally, I would argue for control over whether it becomes a genetically engineered laboratory product or whether it is developed through a harvesting method that keeps economic benefits in the Congo. These are choices that the people of the Congo should be allowed to make, not Monsanto or whoever else applies for the patent on the fungus.

What's the real-world process for these small countries up against powerful corporations?

Very often, these small countries are not fully aware of these laws and considerations. If the company simply sends a negotiator over there and says, This is great stuff, we'll pay you .01 percent of the royalties and we'll build you an airstrip in the community where the fungus was found, and we'll take you to dinner, and so forth. They have indeed, in past deals, bought the proverbial Manhattan for beads. It's only when these countries are fully aware of their options under the law, and hopefully they include those people affected by these decisions so it's not only the higher levels of government making these choices, that we see a close to ideal process. But the legal options in place thus far for developing countries and the general public are better available under the biodiversity treaty, and the U.N. system generally, than under the trade rules, which are strictly commercial and corporate-oriented. The Battle of the Twenty-first Century is going on between trade as an ideology and as a body in international law with clout versus the United Nations, which, despite its problems with clout, has a far more democratic base of laws in place.

Where does morality along the lines of **ordre publique** ***come into legal maneuvering against biotech?***

Ordre publique is a French phrase meaning "public order" that has snuck into a number of international laws. In particular, it was written into the WTO's TRIPs agreement as a loophole. It says that when there is an issue of morality or public order, a government can opt not to follow the TRIPs rules. Again, it's obviously subjective what these panels and the appellate body of the WTO would eventually have to rule on, but let's say the Congo wants to make an issue of this fungus and its development into an insulin substitute. If a drug company in the United States went ahead and took out a patent on it, the Congo could file a complaint under the biodiversity treaty procedure, which simply says that the two parties in a dispute should sit down and talk about it, with an arbitrator if necessary, and settle it. This is how U.N. treaties handle dispute settlement. Under the WTO, the Congolese people and government could say, We think that under the *sui generis* option and under *ordre publique,* we are not compelled to honor a patent. There would be many years of debate to follow through the dispute process. Who knows how it would affect the country in question. In Europe, there are practically riots over the biotech stuff, according to the press. I'm not advocating riots, but sometimes they happen. It's the closest I can get to imagining when a defense on *ordre publique* would actually be defensible. On the other hand, the defense of morality has any number of churches and religious bodies coming out against the patenting of life. There have been quite a few declarations by religious institutions saying it offends their system of belief, and although it is civil, I think it is an example of public order being questioned.

The human rights question related to the patenting of life is actually the basis of a lawsuit in Europe whereby both the Dutch and Italian governments have taken the European Union to the European Court of Justice, arguing that the new European patent law, which does allow for the patenting of life, is illegal and in violation of human rights. The European Court of Justice will now be looking at this question. Its impact on other legal bodies and public interpretation of law will be very interesting.

If U.S. industry and government represents the most aggressive player for GMOs, patents, and so forth on the international level, how do you see American citizens changing that course?

Campaign finance is very important. Public awareness is key, but we have the problem that the media is largely captured by just a handful of corporate interests. Education is key, too, but with the withdrawal of support for public education from the federal and often from our state and local sources, there's been a trend toward corporate funding of education, which in turn leads to a certain amount of bias and influence.

Regarding the WTO, there's very active work going on that will culminate in late November 1999 in Seattle, Washington, at the next meeting of the ministers of the WTO. We expect to see tens of thousands of people from all over the world participating in a protest of WTO policies and its impact on society. Their official goal is to set the agenda for the next several years of the WTO, which means that whether or not our government is able to get their way on biotech won't be decided for the next three or four years, and quite possibly many more if it takes as long as it did last time to reach a final agreement. So, we get quite a few years of campaign time.

How would you encourage people to act?

It's difficult because democracy doesn't extend to the international level, so there aren't any tried-and-true mechanisms. If people came to Seattle to join the gigantic protest—after all, that's what finally led to the end of the Vietnam war—it would be a very significant act of citizenship. Then again, using every avenue possible is important. For example, every time people go grocery shopping, they should ask whether the products they are about to buy are genetically engineered or not. At the moment, store managers won't know, but it will register as a concern. If a lot of people did this, or got up a petition of signatures for the supermarket manager's office, it would make a difference. We are just starting to see here what is happening in Europe, where stores want to meet consumer demand for non–genetically engineered products. We should achieve labeling of our foods at the least. We should insist on prerelease assessment and liability provisions. We should argue antitrust against corporate control of the food system. For example, the U.S. Department of Agriculture tried to pass new regulations in 1998 that would have made the definition of organic foods include genetically engineered foods. Amazingly enough, this proposal generated more letters from the public objecting than any other regulation in the history of the United States. So much decision-making power is at the international level now that the question of global democracy is really at the heart of this whole conversation.

MARTIN TEITEL

(Interview)

FRAMING ETHICAL DEBATES

Casey Walker: Within our lifetime we've witnessed industry shift its base of power from resource extraction to communications/information, and, recently, to genetic information. Each evolution has been publicly sanctioned by assumptions of inevitability and progress toward a better world. Will you begin by critiquing those same assumptions for biotechnology?

Martin Teitel: The idea of progress is a myth, particularly when you apply it to biology. Human beings construct and reconstruct the world according to their own ideas and cultures and agendas, but the biological world works very differently from the world of human ideas. It operates under a different set of assumptions and principles, which is why the term *genetic engineering* is appropriate. We are teleological. We march toward goals that we've decided upon in our minds, and we try to shape and fashion the world to reach those goals. And that's called progress. Putting nature's system of biological processes into an engineering-oriented, teleological-oriented, and progress-oriented framework is, I think, destructive by definition. It's degenerative. I'm not making out some kind of romantic, Rousseauistic case that we should all wear animal skins and run around in the forest. I have my laptop computer and my Dodge van. But there's a big difference between living in and learning from the world, and shaping that world so that it "lives" within our ideas, which is precisely what genetic engineering does.

David Noble's books World Without Women *and* The Religion of Technology *trace historical shifts in religious and political power from the embodied world to the technological world—assumptions so deeply ingrained and part of our milieu that we rarely question them. How do you see the critique of biogenetic engineering becoming proactive along these lines?*

One of the reasons I like to talk about slippery slopes is that we have been sliding on a slippery slope for a long time: the upper end of our slippery slope is the adoption of science as a religion and the consequent impoverishment of our epistemology. At the root of this process, our ways of knowing have become constrained. Those who go outside the approved epistemology are labeled as heretical and are treated the way heretics have been treated for thousands of years, whether it's through the denial of tenure, or the kind of frothing, hysterical editorial about Jeremy Rifkin (author of *The Biotech Century*) that appears in the *Wall Street Journal* from time to time. I call it religion because, ironically, it's based on faith more than on assumptions. This is okay for people to do, I suppose, but it may be the first religion in history with a core built upon the denial it's a religion.

And criticism of it is dismissed as "retro," as Luddite, rather than genuinely turning our attention to the ethical and intellectual debates essential to the world we are creating.

Yes. There are really two things operating here that explain why we're in very, very deep trouble around biotechnology. One is this very thin soup of an epistemology on which we're trying to construct our complicated society. The second is an absolutely astounding release of greed. When we have the coupling of a shaky social and values mechanism with a fierce engine of acquisition and possession, we get something very ugly. And that's what we're seeing in the biotech revolution. For example, in the last twenty-four months, we've seen the utter transformation of the basis of agriculture that has been around for thousands of years. And guess what? Nobody noticed!

I adore imagining conspiracies, and while I think there isn't actually much of a conspiracy out there in the biotech business, I also see that if you were to decide to design a conspiracy to dominate the world, the first thing to do would be to get control of the media. Well, that job has been done quite thoroughly. It is very, very difficult for heretical opinions to be expressed to the general and mainstream public. The next thing to do, if you were designing a worldwide conspiracy, would be to go ahead and take over agriculture, the means of feeding people. Then you would take over the pharmaceutical industry, change the way labor works on this planet, and start owning life forms. In fact, all of these changes are coming about quickly, conspiracy or not. And, since few people know about it, given the media monopoly, what will the public do about it?

Reading polls, even the polls the industry does, we see that when the right questions are put to people—such as "Do you want to know if there are genetically modified organisms in your food?"—an overwhelming, off-the-charts percentage of people say, "Yes." That encourages me. Ironically, it also encourages the biotech industry and the U.S. government, which supports the biotech industry, to do everything they can to maintain universal ignorance about what's going on.

In one of the Council for Responsible Genetics' position papers, there is a critique of professional bioethicists whose job is to make new applications of genetic engineering desirable to the public. Will you describe this growing industry of bioethics and its ties to the media?

Yes. Bioethics is a profession in which people are paid money to render a defensible opinion about the ethics of new developments in biology. Well, he who pays the piper calls the tune. Truly, the question is, Under what auspices is the decision to be rendered? To whom are these people beholden? Who constructs the curricula? I'm not smearing all bioethicists. There are many people working in this area who are utterly sincere, uncorrupt, and independent. Yet one continually encounters people in this business who make pronouncements and turn out to be part of a system in which the subject of their judgment is constrained, a system in which their questions represent a particular point of view. There are some mechanisms, particularly around universities, that permit independent, ethical review of genetic engineering experimentation. I know some people who sit on these review committees for human experimentation and some are wonderful people, yet they have constraints in terms of what they can do. Obviously, it's not only appropriate but necessary for outside people to comment on the process and for independent media such as yours to make sure the public is able to enter into the conversation.

How do we turn attention to the real causes of systemic degradation?

The very ugly word here is commodification. Turning something into a commodity means transforming what we love and care about, and what we connect with in the world around us, into something that is an owned product, something that is fungible, something that is subject to an external control calling itself progress or science or whatever the slogan may be.

Biotechnology is presenting us with a wonderful opportunity right now to ask some deeply reflective questions. What is a human being? What is life? It may appear sophomoric, but these are not theoretical questions, these are real-life, on-the-lab-bench questions of science, and therefore actual questions of public policy, religion, morality, and ethics. We can ask how social agendas are built into biotechnology. We can ask, How can the public see that genetically engineered miracle crops can and are causing starvation? How can the public see and understand that the pursuit of biotech "miracle drugs" is also the pursuit of a particular kind of profit at the expense of some sick people?

There are two approaches. One is that we have to have this kind of conversation loudly and publicly in schools and colleges and on street corners. People have to reclaim the connection to their own bodies, their own biology. Part of this conversation will also consist of reclaiming our language and images. So many of our daily metaphors are coming out of computers rather than out of living things. People are likely to find intimate humor in asides such as, "I woke up this morning with a crashed hard drive and booted up with an espresso." We should be using garden metaphors, not mechanistic language to describe our existence or the world.

The second approach, after deciding to talk with one another about these issues, is to be courageous. Years ago, I listened to a taped, public conversation between Daniel Ellsberg and Ram Dass on how to get rid of nuclear power and nuclear weapons. They speculated that the only way to get rid of them was a catastrophe. They imagined, well in advance, the significance of Three Mile Island and Chernobyl. Here are two extraordinarily decent and thoughtful people saying that maybe the only way humanity will learn is from a catastrophe! As people concerned about biogenetics, we need to have that same conversation. We need to ask about the dangers. What will happen if even a few of our worst fears come true? There are, after all, field trials going on right now that could let loose all kinds of nasty things. A fairly significant amount of work is going on in xenotransplantation, the crossing over of species lines between human beings and animals. These experiments run the risk of releasing pathogens into humanity that can't be stopped. I'm not saying this to be an alarmist—I don't think we serve any good by being alarmist—but how do we serve any good by dismissing dangers and refusing to face the realities?

Biotech does offer us an opportunity to ask, What is a human being? What makes up what we cherish and love? It also poses the opportunity to see how the whole, living system works together.

You've said: "We should never let our rhetoric or our dreams cloud our assessment of the power and strength of our adversaries," and "Corporate accountability and citizen oversight are feasible." How do you see accountability and oversight coming about?

We always have a choice. We have a choice as activists to look at the power of a Monsanto, or a Time-Warner, and say, Here we are, scruffy and shrill, with no hope of obtaining resources that those corporate people command. Let's give it up. Or we can say, Let's be strategic. Let's level the playing field. Let's take these powers on. Our power is in the tremendous array of culture and history and human emotion that have not been commodified and cannot be commanded by our adversaries. The latter gives me immense hope for our ability to get a grip and turn things around. We have to be hard-nosed and practical, and not romanticized by our own rhetoric and siren song. A great model to follow is the Nestle boycott. A group of people were extremely clearheaded and methodical in planning and executing that campaign with great precision and were hugely successful. They stayed clear and focused, and never abandoned their integrity.

Will you speak to various projects, such as HUGO, that are now under way, and an emerging field of the medical industry often referred to as "biopiracy"?

Yes. There are a number of massive projects that deal with population genetics such as HUGO, the Human Genome Diversity Project, and several derivative projects. All of these have certain characteristics. First, there are vast, rather extraordinary amounts of money involved in these projects, which is important to note because money gravitates to the greatest return on an investment. Second, all of these projects have commodification as their main agenda. There is a large amount of life patenting coming out of these projects. This means that while we are constantly told how scientists are doing innocent-sounding things like mapping our genes, they really are mapping in the sense that Columbus or Vasco da Gama mapped. They are methodically charting the territory that they will plant their flags on and make their own. In genetics, this bioinvasion means they are looking to own human genes so they can charge large numbers of us a lot of money for access to products that will make these bioimperialists very, very rich. Third, these projects represent the ultimate in cultural hegemony—that is, a worldview from the rationalist, corporatized West that is being imposed on all of humanity at the genetic level. Finally, and particularly with the new environmental genome project, there is an agenda that appears to be eugenic in nature—that is, to wield ultimate power on the level of population genetics to redesign various physical traits and outcomes. In the real world this means gaining control of the human germline, the literal basis for who we are. Then these fellows, who evidently imagine they have the right to do this, will change humanity. They want to make changes that will be inherited, so that all succeeding generations of people fit not nature's design or God's design or what have you, but the design that emanates from the values of these almost unimaginably hubristic scientists.

Much of this science makes judgments about individuals: who gets to be born and who doesn't and therefore what humanity looks like and consists of. Sometimes, when I'm speaking in public, scientists become irritated with me and say, You don't have the credentials to speak about science. You're not a scientist. And I say, Fine. If we're going to have the same ground rules, then you don't have the credentials to make judgments about moral, ethical, and spiritual matters. Show me how that was part of your training and background. Will you stay the heck out of those areas? Of course, if they did, they would have to close their labs. The very concept of genetic engineering science is totally infiltrated with a set of covert values and moral schema. These large-scale scientific/economic projects, carried out in public but in secret, imply some of the most extraordinary arrogance since Tamerlane or Genghis Khan.

The issue of biotech industry contracts with universities sets the deeper question we began with, which is, What kind of society are we creating? If it is common knowledge that today's educational institutions are deeply invested, literally and pedagogically, in the technical and commercial application of biological knowledge, where is public comment on this?

A respected professor is just now finishing research to document conflicts of interest between academia and industry. Even he, who designed this research, was stunned by the amount of conflict there is. Industry asked the same question you did, and they said, Let's get in there and get a piece of that action and make sure that we're in charge of what's happening.

My kids go to public schools in suburban Boston that are considered excellent, but I can say these schools are shockingly deficient. They are, basically, vocational schools. There's too little in them that I would call a decent liberal arts training. They're purely aimed at preparing children for the marketplace. The real liberal arts deficiency extends from questions of values, to questions of curriculum content, to homework assignments. The economic worldview permeates, saturates, everything that happens in these children's experience. We're turning out little worker bees, not engaged citizens, not thoughtful neighbors, and not loving human beings.

Will you speak to the controversies associated with DNA databanks?

Our government is very busily eroding our privacy on a genetic level with the proliferation of DNA databanks, some of which they mandate. The game is not only not over for those of us who want protection of our privacy, but it has gone far in the wrong direction, which is just now beginning to penetrate public consciousness. I'm receiving a lot of requests for information and interviews on DNA databanks. People ask, Isn't this a way to catch criminals? But this question is the criminal justice system's version of "we will feed the hungry" or "we will cure the sick"—myths sold by the biotech industry and their friends in government. They say, We will catch the bad people. How can you be against that? Phil Bereano, at the University of Washington and on the board of the ACLU, makes the point that we could also catch more bad people if we would say it's okay for the police to kick down our doors at will, or to stop us on the street and search us, or to open our mail. Are we willing to put up with absolutely anything to be more secure? Since the answer is no, we won't accept just any violation of our rights for the illusion or the actuality of greater security, then where do we draw the line? We have the Bill of Rights and the ACLU to help us figure out where to draw that line, and we ignore them at our peril.

As a society, we have fought very hard with very good reason to have certain safeguards for privacy and individual boundaries, and to constrain the criminal justice system. Now they're invading our bodies and our reproductive potential: our genes. The Fourth Amendment, which covers search and seizure, has been interpreted in many court cases as stopping at the skin. That's why fingerprints are permitted, even though they technically violate the Fourth Amendment. Specifically, fingerprints are allowed because they're an image of the outside of your body. The case law on this is amazingly explicit. But now there are DNA databanks. The last bastion of government resistance was the Massachusetts Supreme Court, and I happened to go and watch the court session on the case that now basically takes away the ability of certain citizens to keep the insides of their body private from their government. People argue with me, Well, we want to catch the bad guys. Okay! But we are also rapidly sliding down a very slippery slope.

Will you comment on the now famous case of Mr. Moore's spleen, in which he lost rights to his spleen tissue because it had been removed from his body during surgery, and therefore lost any claim to its estimated one billion dollars' worth of derivative protein profit going to the University of California?

Yes, his doctor did real well. It's worth noting, though, that John Moore's claim was not privacy, it was theft. In other words, John Moore did not raise the issue that it's just plain wrong to do what was done to him. He wanted a piece of the action, saying his doctor had no right to alienate his body's tissues and not give him a share of the profits. I would raise different issues. John Moore is the person who was put through that horrible experience and I'm not making a judgment about him. But if they were, to use his phrase, "stealing" parts of my body, I'd want to talk about the right of another human being to do that. It's a very personal violation to take what is most essentially me and make it the property of another human being. In my theology, that's one definition of sin. I'd also want to talk about a supposed health care system that permits my doctor to take ownership of my body parts, or a government that allows and encourages companies to assert that they own life. Talk about sin.

People argue back and forth about organ transplants and different species' tissues being introduced to save or prolong life. Where do you make ethical decisions in this area?

I'm glad you asked that, because we're letting industry, with its particular agenda of profit, define the questions. They say, correctly, that 20 percent of people in the United States waiting for an organ transplant die before an organ becomes available. They follow that statement with, How can anyone be against xenotransplantation, or against other technologies that will save these lives? The answer is that there are more questions to be asked and possible answers to be found. Context matters, particularly the financial context—the profits at stake—for the person framing the ethical question about who gets organs and lives, and who doesn't get organs and dies.

Here are two examples of larger contexts. First, in Spain, virtually no one dies awaiting transplants because Spanish law maintains presumptive consent. You must affirmatively opt out of the transplant system if you don't want to have your organs donated after you die. In our country, you have to affirmatively opt in. Civil libertarians may take issue with presumptive consent, and surely it needs to be thought through, but organ availability was a policy decision, not a science decision, that solved transplant shortages in Spain. We have not had this conversation in our country—the issue hasn't been raised here. We haven't been given the opportunity to sit down with civil libertarians and say, What about presumptive consent? Is that an invasion? It's a solution that hasn't been put into the debate on xenotransplantation, though I assure you xenotransplantation experts know about it.

Second, there are people in the biotech industry talking about genetically engineering cows' milk so that it will resemble and replace human milk. They argue that women who are concerned about their breast milk being contaminated by PCBs or PBBs or other environmental toxins out there can now feed their babies genetically engineered cows' milk that doesn't contain the nasty pollutants that could hurt their babies.

What about localized antibodies, and what about the bonding that occurs with breast feeding!?

Yes, there's a long list of "what abouts"! One of the "what abouts" is asking if any of the companies suggesting genetically engineered cows' milk are the same companies that have perpetrated the pollution that makes women's breast milk unsafe? Why not address the problem of polluted breast milk by dealing with the pollution? The reason is, of course, that businesses don't profit from reducing pollution. Over and over again, the wrong questions are asked and the wrong selection of solutions is presented to the public, to consumers. This is one of the primary problems activists have to address: reframing the questions so we are pursuing the right kinds of answers. Let's not buy into living in a world of narrow choices and debates that are framed by corporations and constrained by the images projected by well-paid public relations firms.

Will you describe the situation you've cited in Council for Responsible Genetics literature—that there have been two hundred documented cases of discrimination based on preexisting genetic conditions? Are we heading pell-mell into a society of "have and have-not genes"?

The figure of two hundred cases in our files refers only to our very tiny sample of reality. Society makes labels, and from those labels makes decisions. The real basis of discrimination is that certain characteristics are held to be a problem that should be tested for, and if it's a prenatal situation, then we are told to eliminate it; if it's a preemployment situation, we don't hire. One of the many problems with permitting genetic discrimination on the basis of labels is that we have no control over what will be the label du jour.

Right now we talk about people who have a certain kind of condition that we define as an "illness." People in the disability community often don't appreciate seeing their lives defined by what are seen by other people as limitations, much less hearing the medical establishment say, These are people who never should have been born, and we can help you to prevent more of these people from coming into being. Imagine the rich, full life of a person being reduced to one characteristic that is labeled, by powerful people external to them, as reason that they never should have been born! Furthermore, how well will society accept and support people whom they are told should never have been born in the first place?

The top of the slippery slope is preventing more cases of Tay-Sachs disease and the bottom of the slippery slope is "ethnic cleansing," or Kosovo. I've done a great deal of human rights work in my life, and a common denominator in many situations ranging from, say, South Africa to China is that they define the person they're about to oppress as differently human, and then subhuman. Human rights are those rights that accrue to anyone who is defined as human. Whether it's Kosovo or South African apartheid or a dictatorship in Chile, if you read the rhetoric of oppression, it's amazingly similar. The oppressors say those people are not quite human, not quite worthy in the same way as the ones generating the rhetoric.

Today we can see a situation of genetic apartheid, in which people are defined by some in the medical establishment who use unbelievably condescending rhetoric. "Genetic discrimination" is too gentle a term for the harm it does our sisters and brothers.

One of the values of the deep ecology movement is that it recognizes intrinsic rights and values across the board. In a world using deep ecology morality, we would no longer look at the value of a cow or a pig, or even an ear of corn, as merely valuable in terms of its usefulness.

Yes. I like to go past neutral terms, especially one like "rights"—which connects me with legal things and makes me shiver—to instead say "worth." It has to be okay for rational people in our society to talk in terms of value judgments, for that to be part of polite discourse instead of this pseudoscientific sham that we're all going to be legalistic and scientifically neutral—since we're not. It isn't just that another human being across the street or an ear of corn is okay, but that they have value, and that my connectedness to that other living thing is not neutral but an affiliation, an attraction to those things as other living beings in my world.

Beauty, too, is a much stronger force in our world than rights or even some kind of measurement of worth. Children all over are asking their parents, What about the shootings in these schools? What about Kosovo? It's important for adults to be able to give authentic responses to these kinds of questions. One of the things we can say is: When a mother who is wheeling her child in a carriage down the street stops and leans down and looks deeply into that child's face and touches its cheek, it doesn't make the six o'clock news. But those acts of deep love are continual among human beings and from human beings outward in the world—utterly continual and real and strong and present. They're not reported on "Dateline," or even in your fine publication. It's not how we experience the world and what we point at. But actually—and I don't think I'm a Pollyanna—we live in a world that is saturated in kindness and goodness that becomes invisible because we don't attend to it. We don't see it in the same way that we don't count the leaves on a tree. It's there and it's real. And by the way, it's something that can't be commodified, patented, or owned.

DAVID PETERSEN

OF GADGETEERS AND BIONIC DEER: THE BIOTECH BASTARDIZATION OF HUNTING

For preagricultural, foraging peoples—our "savage" human forebears—sacred and secular were inseparable. The same wild animals preyed upon were prayed to, and with life and death the yin and yang of daily existence, sacredness was diffused throughout.

It's still true today, of course, that life feeds on death. But unlike our ancestors, who knew exactly where their meat came from and at exactly what costs, we modern carnivores (most of us, most of the time) buy our meat as deanimated product without considering that it was in fact quite recently part of a living, albeit genetically deracinated and chemically larded, animal. This commercial distancing from our food makes it easy for us to euphemize, ignore, and, in extreme cases such as veganism, deny the deaths that nourish all our lives.

Yet for an open-eyed minority of modern hunters, the ancient animistic sense of soulful, visceral unity with wildness—diffused sacredness, if you will—remains grounded as ever in awe, humility, and reciprocity. "Wildness is what I kill and eat," proclaims Paul Shepard, the father of human ecology, "because I too am wild."

Evolutionarily, hunting is a definitive human activity. The hunting/gathering lifeway—owing to the distinctive ways of living, thinking, and worshiping it fostered—was the prime mover behind our becoming human. As Paul Shepard testifies, "The dynamic of escape and pursuit is the great sculptor of brains." Spiritually, when we "lived" wild animals daily, we came to think, pray, and be as wild animals.

Thus did hunting patiently guide our formative humanity toward a universal animistic (zoomorphic) cosmology, binding humanity intimately to earth via the great gastronomical round of life-giving death.

For at least the final 1.8 million years of human evolution—coincident with the emergence of the penultimate hominid, *Homo erectus*—this was the way life was, the way we were. By 50,000 to 100,000 years ago, near the apogee of the icy upper Pleistocene—The Age of Man—*Homo sap* was a done deal. Thus, when Paul Shepard declares that "wilderness is where my genome lives," he reminds us that genetically, socially, physiologically, psychologically, and spiritually we exist today as "space-needing, wild-country Pleistocene beings trapped in overdense numbers in devastated, simplified ecosystems."

Shepard is hardly alone among scientific philosophers in proposing that this gaping mismatch between our lingering Pleistocene design for wildness and our current concrete culture is responsible for much or most of the angst, immaturity, violence, environmental genocide, and multifaceted suffering plaguing the world today.

Which is not meant to suggest that hunting is a universal antidote to our pandemic existential despair, being neither right for nor available to everyone. Yet neoanimistically informed hunting, because phylogenetically it remains perhaps the definitively human activity, continues to provide a reliable and wholly organic pathway to what Shepard would term *phylogenetic felicity.* But only if you do it right. And there's the rub.

After five thousand years of civilization, with its acutely humanized morality and absurdly alien spiritual paradigms, many if not most modern sportsmen have lost contact with the ancient gossamer thread twining hunting with spirit. No longer do they enter nature in pursuit of spiritual as well as physical sustenance. Rather, most hunters today seem motivated by trivial, pointless, morally questionable desires; by an aching boredom and consequent hunger for escapist divertissement. Thus, modern hunting's problem sprouts not from hunting per se but from the current culture of hunting. And that culture is shaped, which is to say misshaped, by the commercial world in which, to one degree or another, we all must live, work, worship, and play.

"What can be said about hunting," asks Edward Abbey, "that hasn't been said before? Such a storm of conflicting emotions!"

Indeed, amid all the confusion and controversy surrounding hunting today, what can be said with objective authority about hunting and hunters, their worldviews, motivations, and morals?

First and foremost, we can say that modern hunters are far more dissimilar than they are similar, and that the concept of a cohesive hunting community is flatly bogus. A quarter-century ago, Yale research professor Stephen R. Kellert conducted a study involving some 3,200 participants nationwide—a study that, even today, remains the definitive academic statement on the attitudes of hunters (and antihunters) toward wildlife.

Based on mountainous data, Kellert split hunters into three broad groups: utilitarian/meat hunters (43.8 percent of those who had hunted within the previous five years), dominionistic/sport hunters (38.5 percent), and naturalistic/nature hunters (17.7 percent).

As their title suggests, utilitarian/meat hunters claimed to hunt only for food. Almost exclusively male, they were older and lower in education and income than the national average. In addition, most had rural agricultural backgrounds. Although meat hunters tested fairly well on Kellert's animal-knowledge scale, they registered coldly utilitarian attitudes toward animals, as reflected in their support of trapping, predator control, and other "practical uses" of wildlife. It's a type, Dr. Kellert and I agree, that has dwindled nationwide as a result of increased urbanization, education, and income.

By contrast, the dominionistic/sports were dominantly urban and knew little about nature. To this group, says Kellert, "the hunted animal was valued largely for the opportunities it provided to engage in a sporting activity involving mastery, competition, shooting skill and expressions of prowess." Whether the dominionistic/sport subset has grown or shrunk in the quarter-century since Kellert's study, it certainly has become more visible, more technologically oriented, and more materially affluent.

Stephen Kellert was still a boy when pioneering hunter/conservationist Aldo Leopold first recognized the fathers and grandfathers of Kellert's dominionistic/sport set. After praising the skills, independence, effort, humility, and naturalistic outlook inherent to traditional hunting, Leopold went on to damn the hunters: "And then came the gadgeteer, otherwise known as the sporting-goods dealer. He has draped the American outdoorsman with an infinity of contraptions, all offered as aids to self-reliance, hardihood, woodcraft, or marksmanship, but too often functioning as substitutes for them. Gadgets fill the pockets, they dangle from neck and belt. . . . I have the impression the American sportsman is puzzled; he doesn't understand what is happening to him. . . . It has not dawned on him that the outdoor recreations are essentially primitive; atavistic; that their value is a contrast-value. . . . The sportsman has no leaders to tell him what is wrong. The sporting press no longer represents sport; it has turned billboard for the gadgeteer."

Today, not only the sporting press, but the Internet has turned "billboard for the gadgeteer." Consider the fall 1998 issue of *HuntingNet* magazine: "The official publication of the world's largest hunting website." Announcing the issue's theme is a bold, gold headline, writ large across the chest of a young, Ramboesque cover model: "Takin' the High-Tech Road." Gadgets fill the pockets of this virtual bowhunter's computer-designed camouflage clothing and, indeed, dangle from neck and belt. To hammer the point home, each of the goodies is highlighted in a close-up captioned photo: GPS, mechanical broadhead, night-vision binoculars, electronic range-finder.

To take maximum advantage of most states' laughably liberal interpretation of primitive weapons—users of which are granted longer and otherwise more favorable hunting seasons—our synthetic nimrod is equipped with a space-age compound bow constructed almost entirely of high-tech synthetics and elaborately configured with cams, cables, pulleys, overdraw, sight, stabilizer, and other modern "primitive hunting" technocrutches.

In Nature Boy's "free" hand, he carries a portable self-climbing tree stand—an essential aid for outwitting those clever suburban Bambis and Falines. But for all of this, where, we're left to wonder, is Gadgetman's infrared heat-source detector, his bionic ear, game-trail timer, two-way radio, mechanical string release, chemical odor-eliminator scent-proof suit, synthetic rattling antlers, and all the other junk that's routinely hawked and editorially hyped in the commercially venal sporting press today? Surely, no self-respecting primitive-weapons hunter would venture afield without such traditional essentials as these. Is our boy perhaps underequipped? More likely, all that unseen booty is stashed in his ATV, waiting on a trailer behind his SUV, over on the far side of the cornfield.

Nor is gadget-mania the worst of it. What, I wonder, would old Aldo think of the current cult who pay to kill captive, increasingly biogenetically engineered trophy animals?

This sick little story begins, and ends, with the worst and least of the testosterone-drenched subset Stephen Kellert calls dominionistic/sports. This lunatic fringe—a tiny minority even among the headhunting crowd—literally buy their "trophies of a lifetime," paying big bucks to indulge in the shooting-gallery executions of big bucks. And increasingly, the victims in this bloody for-profit business—trophy deer, elk, and other "game" species—are selectively bred for specific morphological traits—big horns, antlers, or skulls. Many have pet names and are tame enough to eat from your hand.

By definition, any "hunt" requiring no hunting at all is no hunt at all. Yet the privatization and genetic manipulation of wildlife for profit, euphemized as "alternative livestock ranching" (or more de-animating yet, as "farming"), is booming in the rural western United States and Canada. Moreover, it's eagerly endorsed by most state and provincial departments of agriculture. Never mind that game farming and canned killing are roundly decried everywhere by concerned wildlife managers, ethical hunters, and those relatively few among the non-hunting public who know of it and give a damn.

How can this be? Why is canned wildlife killing so ubiquitously legal?

Landowners' rights. States' rights. Culturally inculcated Cartesian dualism. Money. Jobs. Politics. The American way.

And please remember this: Each time you enjoy a meal of "wild game" at a restaurant, you become an active participant in the obscene cruelty of game ranching. We have enough domesticated "genetic goofies" already, let wildlife remain wild.

To date, the blight of bioengineering has yet to infect public wildlife or true hunting; no designer-breeding of trophy-antlered elk or deer for release into the wild. So far, the biomanipulation of wildlife remains limited to the private arena: hybridizing pen-raised "hunting preserve" pheasants to maximize their "sporting qualities"; transplanting embryos from one subspecies of elk into another in hopes of building a super-subspecies; feeding nutritionally hot supplements to promote unnatural antler growth; sperm-banking and artificial insemination of "factory mother" cervids with the semen of trophy males.

Regarding the latter caper, the stellar example is "30–30," the world's largest-antlered captive whitetail buck. In 1996, 30–30 was purchased at stud for $150,000. Today, his semen is worth more than gold-plated cocaine. After being electroejaculated, each load is divvied into several dozen test tube doses that sell on a seller's market for $1,500 a squirt.

Which brings us back around to those smarmy little cowards who finance this Nazi Dr. Doolittle circus of horrors, gleefully forking over $8,500 for the privilege of gunning down a custom-bred, alfalfa-fattened, trophy deer. Execution rights to a big bull elk can cost a whole lot more.

Delighted by all of this and dreaming of an ever more profitable future, green-eyed wildlife geneticists, according to an article in *Sports Afield,* "are working to map out the genetic code of whitetails so they can isolate the antler chromosome that will make genetic engineering for big-racked deer available in the near future. With that map will come the possibility of cloned trophy deer."

Yes, and given that brave new biotech breakthrough, why limit our God-playing to the private fenced pasture, where only the filthy rich can participate and profit? Already, some quality-conscious public-lands head-hunters are petitioning state and provincial wildlife agencies to adapt the wonders of biobastardizing to the genetic enhancement of public wildlife on public lands.

Should this ever happen, true hunting is dead. Naturally evolved wildness, likewise, will be no more. And where to then? "A world of made," counsels e.e. cummings, "is not a world of born."

All cultures are made. And ours is made to worship efficiency: fast, easy, and certain. To transport this workaday, time-clock mentality into what should be the challenging, meditative, and magically uncertain adventure of the hunt, is to trivialize one of humanity's oldest, most rewarding, joyful, and, for those of us blessed and cursed with hunters' hearts, sacred acts.

This is modern dominionistic/sport hunting's central problem—a collective failure of the spirit, precipitating the lockstep erosion of both internal ethics and external respect.

Yet contrary to what myopic or disingenuous antihunters would have the public believe, all is not doom and gloom in the world of modern hunting. "In defiance of mass culture," Paul Shepard proclaims, "tribalism constantly resurfaces." And he is right. Standing in proud contrast to the egoistic sport and the pragmatically utilitarian hunter, is Kellert's third type, the naturalistic/nature hunter. As a group, Kellert's nature hunters were younger, more educated and affluent than meat or sport hunters. This tribe, not surprisingly, also included the highest percentage of female hunters. Further distinguishing themselves, Kellert's nature hunters participated not only in hunting but in such "nonconsumptive" outdoor activities as camping, backpacking, and bird-watching. (In an ironic confluence of opposites, these same passions, minus hunting of course, were shared by many of the most ardent antihunters in Kellert's study.)

Nature hunters also hunted more often than members of the other two groups, "perhaps suggesting a stronger commitment . . . to the activity." Moreover and significantly, nature hunters scored the highest "knowledge-of-animals scale scores" of all those tested—hunters, nonhunters, and antihunters alike. While the dominionistic/sport staggers aimlessly through his gadget-confined world of made, and the meat hunter thinks only of his stomach, the nature hunter is an active and versatile player in the big wide world of born. He or she is also a de facto neo-animist, a bona fide spiritualist (whether she or he knows it or not), perpetuating the bottomless tradition of our prelapsarian ancestors, the archetypal nature hunters.

Whether ancient or neo, the animistic worldview, as defined by anthropologist Richard Nelson, embraces all of nature as "spiritual, conscious and subject to rules of respectful behavior." It's hardly coincidental, then, that nature hunters—including the likes of Richard Nelson, Aldo Leopold, Michael Soule, Dave Foreman, and myriad notable others—so often number among our most passionate nature lovers and defenders.

In dominionistic/sport hunting, with its gadget-addiction and lust for "bigger and better," even unto the extremes of bioengineering, the traditional hunter's heritage of animistic altruism is not only lost, it's openly mocked. For hunting to survive, for hunting to deserve to survive, this must change.

Ironically, the antihunting movement is the least likely tool for affecting needed hunting reform—insofar as most animal rights champions (Kellert's two basic types are "humanistic" and "moralistic") are driven more by an emotionalized fervor to censor others than by an informed biologic, and appear incapable of distinguishing nature hunters from canned killers.

A better bet is informed and selective criticism of contemporary hunting values and practices, sounded from within as well as without the hunting ranks. "To criticize the bad," Ed Abbey reminds us, "is our duty to the good." And among the baddest of the bad today, in hunting as elsewhere, is biotechnology. At the very least, therefore, all who care—authentic hunters, concerned nonhunters, and antihunters alike—must find a way to unite in condemning biotechnology for what it is: a super-weapon in befuddled humanity's war against natural-born wildness . . . and thus, against ourselves.

About the Contributors

Wendell Berry is the well-known author of several dozen books of poetry, fiction, and essays, including *What Are People For?, The Gift of Good Land, Home Economics, Standing by Words, The Unsettling of America, Fidelity, Sabbath Poems,* and *The Timbered Choir.* He has received numerous awards for his work, including the T.S. Eliot Award, the Aiken Taylor Award for Poetry, and the John Hay Award of the Orion Society. He and his wife, Tanya, live on a farm in an agricultural community in Henry County, Kentucky.

Marti Crouch, associate professor of biology, Bloomington, Indiana. Professor Crouch was named one of three winners of the annual Joe A. Callaway award for civic courage from the Shafeek Nader Trust for the Community Interest, 1999.

For a more detailed explanation of the patent referred to in Professor Crouch's contribution, "How to be Prophetic," see "How the Terminator terminates: An Explanation for the Non-Scientist of a Remarkable Patent for Killing Second Generation Seeds of Crop Plants," published as an occasional paper of the Edmonds Institute, 20319 92nd Avenue West, Edmonds, WA 98020, USA. It is also on the web at: http://www.bio.indiana.edu/people/terminator/html.

Kristin Dawkins is the director of the trade and agriculture program within the Institute for Agriculture and Trade Policy. Her work has focused on food security, environmental impacts, and intellectual property rights. She represents the institute at numerous international negotiations and conferences, and writes articles for journals throughout the United States, France, Britain, Germany, Brazil, Malaysia, and South Africa. Dawkins came to IATP from the Harvard Law School Program on Negotiation, where she was senior writer for their international publication, *Consensus.* Earlier work emphasized civic values and the role of citizens in policy formulation.

Chris Desser currently coordinates the Funders Working Group on Biotechnology, an international effort among foundations and activists to raise awareness and activism on health and policy issues concerning transgenics, terminator technologies, and genetically modified foods. In 1999, California governor Gray Davis appointed her to the Coastal Commission; she also serves as the project director of the Migratory Species Project. She was executive director of Earth Day 1990, the largest global demonstration in history, involving over 200 million people in more than 140 countries. For years, Chris practiced environmental law, working in the private, nonprofit, and public sectors. She is a member of the International Forum on Globalization and has served as an adviser or board director for numerous environmental and progressive organizations. For Funders Working Group on Biotechnology information: 415-561-2626

Richard Hayes is director of the Exploratory Initiative on the New Human Genetic Technologies. For the past year he has been writing and speaking about the social and political implications of these technologies, and of the need for public awareness and activism to prevent their misuse. Hayes is a doctoral candidate in energy and resources at UC Berkeley and is a member of the human genetics committee of the Council for Responsible Genetics. He has long been active as an organizer in progressive social and political movements, serving most recently as chair of the Sierra Club's global warming campaign committee, and as assistant political director and director of volunteer development on the national Sierra Club staff. Contact information: rhayes@publicmediacenter.org

Freeman House co-founded the Mattole Watershed Salmon Support Group and the Mattole Restoration Council in the Mattole River valley on the northern coast of California. Known by many for a lifetime of bioregional study, writing, editing, and activism, House's new book *Totem Salmon* (Boston: Beacon Press, 1999) tells of the hard, patient work too full of life to be reduced by today's environmental parlance as "species restoration." *Totem Salmon* is winner of BABRA's (Bay Area Book Reviewers Association) best nonfiction by a northern California writer.

Catherine Keller is the author of *From a Broken Web: Separation, Sexism and Self* and *Apocalypse Now and Then: A Feminist Guide to the End of the World.* She is a professor of constructive theology at Drew University and works in the theories and practices of ecofeminist spirituality and philosophical theology. Currently, she is writing *The Face of the Deep,* on the lost chaos of the creation.

Andrew Kimbrell is a public interest attorney, activist, and author. After eight years as policy director at the Foundation for Economic Trends, Kimbrell established the International Center for Technology Assessment in Washington, D.C., in 1994, and the Center for Food Safety in 1998. His books include *The Human Body Shop: The Cloning, Engineering, and Marketing of Life* and *The Masculine Mystique: The Politics of Masculinity* (1993). He is well known for key litigation, legislation, and media spokesmanship on issues pertaining to technology and human and environmental health. *Utne Reader* named Kimbrell one of the world's leading one hundred visionaries.

David R. Loy is professor in the Faculty of International Studies at Bunkyo University, Chigasaki, Japan. He works primarily in comparative philosophy and religion, particularly with comparing Buddhist with modern Western thought. He is the author of *Nonduality: A Study in Comparative Philosophy* and *Lack and Transcendence: The Problem of Death and Life in Psychotherapy, Existentialism and Buddhism;* and the editor of *Healing Deconstruction: Postmodern Thought in Buddhism and Christianity.*

Stuart Newman is a professor of cell biology and anatomy at New York Medical College, Valhalla, New York; he has degrees in chemistry from Columbia University (A.B.) and the University of Chicago (Ph.D.). He directs a federally funded laboratory in developmental biology and has contributed to several scientific fields, including cell differentiation, theory of biochemical networks and cell pattern formation, protein folding and assembly, and mechanisms of morphological evolution. He is a founding member of the Council for Responsible Genetics, a public interest organization against the misuse of biological science and technology. He has also written articles on the cultural background and social implications of biological research (see "Carnal Boundaries" in *Reinventing Biology: Respect for Life and the Creation of Knowledge,* Lynda Birke and Ruth Hubbard, Editors [Bloomington and Indianapolis: Indiana University Press, 1995]). Dr. Newman has served as a visiting scientist in France and India, and as a consultant to the U.S. National Institutes of Health.

David Petersen lives in the San Juan Mountains of western Colorado. He is the editor of A. B. Guthrie's environmental essays, Edward Abbey's journals, and *A Hunter's Heart,* a thoughtful exploration of hunting ethics by forty-one writers. Petersen is the author of six books on the natural world including *Elkheart: A Personal Tribute to Wapiti and Their World; The Nearby Faraway: A Personal Journey Through the Heart of* *the West; Ghost Grizzlies: Does the Great Bear Still Haunt Colorado?;* and *Heartsblood.*

Richard Strohman is emeritus professor of molecular and cell biology at UC Berkeley; with a Ph.D. from Columbia University. On leave from UCB in 1990, he was research director for the Muscular Dystrophy Association's international effort to combat genetic neuromuscular diseases; earlier he co-authored *Gene Expression in Muscle.* During 1992–93, he was Distinguished Wellness Lecturer at UCB. He continues to teach courses and publish articles, and is currently writing a book on the growing crisis in theoretical biology. He has received wide acclaim as one of the leading figures thinking and writing on opportunities for a new, holistic scientific theory of living systems.

Martin Teitel is executive director of the Council for Responsible Genetics, a national citizens' organization that seeks to increase public participation in decisions about biotechnology and genetic engineering; and editor of *GeneWatch,* a journal of biotechnology activism. He is the author of *Rain Forest in Your Kitchen: The Hidden Connection Between Extinction and Your Supermarket;* has co-authored with Hope Shand *The Ownership of Life: When Patents and Values Clash* and with Kimberly A. Wilson *Genetically Engineered Food: Changing the Nature of Nature.* Contacts for CRG: crg@gene-watch.org; www.gene-watch.org

Jack Turner is the author of *The Abstract Wild* and *Teewinot: A Year in the Teton Range.* A former academic philosopher, Turner devoted much of his life to travel throughout the Himalayas and South America. He is a guide for Exum in Teton National Park, where he has climbed and lived for nearly thirty years.

About the Editor

Casey Walker founded *Wild Duck Review* in 1994 and has edited and published nineteen issues to date, featuring essays, poetry, book reviews, memoirs, and over eighty interviews with literary artists, scientists, cultural critics, activists, and political leaders. She was educated at UC Davis and the Institute for European Studies in Vienna, Austria, in international relations: western European history, with graduate studies in English literature: fiction writing.

Wild Duck Review: Literature, Necessary Mischief, & News was founded in 1994 by editor and publisher Casey Walker and named after a passage from H. D. Thoreau's essay "Walking": "In literature it is only the wild that attracts us. Dullness is but another name for tameness. It is the free and wild thinking in *Hamlet* and the *Iliad,* in all the scriptures and mythologies, not learned in the schools, that delights us. As the wild duck is more swift and beautiful than the tame, so is the wild—the mallard—thought."

Each edition of *Wild Duck Review* takes up contemporary issues, such as the state of media, education, population, biotechnology, or political will, under the intellectual rubric of long-term cultural and ecological vibrancy. Each includes interviews, essays, poetry, indexed centerfolds, and book reviews by leading intellectuals, literary artists, scientists, and political activists—just as those collected in this book.

For a sample copy ($4), back-issue list, or to subscribe ($24/four issues or $32 U.S. outside United States), please contact:

Casey@WildDuckReview.com
www.WildDuckReview.com
530-478-0134
Wild Duck Review
P.O. Box 388
Nevada City, CA 95959-0388